THE OPERATIONS PROCESS

July 2019

United States Government
US Army

Contents

		Page
	PREFACE	iii
	INTRODUCTION	v
Chapter 1	**FUNDAMENTALS OF THE OPERATIONS PROCESS**	**1-1**
	The Nature of Operations	1-1
	Unified Land Operations	1-2
	Mission Command	1-3
	The Framework of the Operations Process	1-4
	Principles of the Operations Process	1-7
	Integrating Processes	1-15
	Battle Rhythm	1-17
Chapter 2	**PLANNING**	**2-1**
	Fundamentals of Planning	2-1
	The Science and Art of Planning	2-2
	The Functions of Planning	2-3
	Planning and the Levels of Warfare	2-7
	Operational Art	2-10
	Integrated Planning	2-15
	Key Components of a Plan	2-19
	Guides to Effective Planning	2-21
	Planning Pitfalls	2-25
Chapter 3	**PREPARATION**	**3-1**
	Fundamentals of Preparation	3-1
	Guides to Effective Preparation	3-2
	Preparation Activities	3-4
Chapter 4	**EXECUTION**	**4-1**
	Fundamentals of Execution	4-1
	Guides to Effective Execution	4-1
	Responsibilities During Execution	4-4
	Execution Activities	4-5
	Rapid Decision-Making and Synchronization Process	4-9
Chapter 5	**ASSESSMENT**	**5-1**
	Fundamentals of Assessment	5-1
	Assessment Activities	5-2

DISTRIBUTION RESTRICTION: This manual is approved for public release; distribution is unlimited.

*This publication supersedes ADP 5-0, dated 17 May 2012, and ADRP 5-0, dated 17 May 2012.

Assessment Process 5-4
Guides to Effective Assessment 5-6

SOURCE NOTES .... Source Notes-1

GLOSSARY Glossary-1

REFERENCES References-1

INDEX Index-1

Figures

Introduction figure-1. Operations process logic chart vi
Figure 1-1. The operations process 1-4
Figure 1-2. The commander's role in the operations process 1-8
Figure 1-3. Commander's visualization 1-9
Figure 2-1. Levels of warfare 2-9
Figure 2-2. Sample line of operations and line of effort 2-13
Figure 2-3. Integrated planning 2-16
Figure 2-4. Activities of Army design methodology 2-17
Figure 3-1. Transition among the integrating cells 3-8
Figure 4-1. Risk reduction factors 4-3
Figure 4-2. Decision making during execution 4-6
Figure 4-3. Rapid decision-making and synchronization process 4-9
Figure 5-1. Activities of assessment 5-2

Tables

Introduction table-1. New, modified, and removed Army terms vii
Table 3-1. Preparation activities 3-4
Table 4-1. Decision types and related actions 4-7

Vignettes

Agility: Rapidly Turning the Third Army to Bastogne 1-5
Collaboration: Meade's Council of War 1-14
Tenets in Action: OPERATION JUST CAUSE 2-22
Prepare: Rangers Train for Seizing Pointe du Hoc 3-3
Large-Unit Preparation: Third Army Readies for OPERATION IRAQI FREEDOM 3-9
Decision Making During Execution: Chamberlain at Little Round Top 4-8
Measures of Effectiveness: OPERATION SUPPORT HOPE 5-4
Commander's Assessment: Are We Ready To Close on Baghdad 5-7

Preface

ADP 5-0 provides doctrine on the operations process. It describes fundamentals for effective planning, preparing, executing, and assessing operations. It describes how commanders, supported by their staffs, employ the operations process to understand situations, make decisions, direct action, and lead forces to mission accomplishment.

To comprehend doctrine contained in ADP 5-0, readers should first understand the fundamentals of unified land operations described in ADP 3-0. As the operations process is the framework for the exercise of command and control, readers should also understand the fundamentals of command and control and mission command described in ADP 6-0. Readers must also understand how the Army ethic guides decision making throughout the operations process (see Army doctrine on the Army profession).

Several publications support ADP 5-0. For detailed tactics and procedures associated with the operations process, such as the duties and responsibilities of the staff, how to conduct the military decision-making process, and formats for plans and orders, readers should refer to FM 6-0. Techniques for organizing command posts and command post operations is located in ATP 6-0.5. Techniques for employing the Army design methodology is located in ATP 5-0.1. Techniques for assessing operations is located in ATP 5-0.3.

The principal audience for ADP 5-0 is Army commanders, leaders, and unit staffs. This publication also provides the foundation for Army training and education curricula on the operations process. Commanders and staffs of Army headquarters that require joint capabilities or form the core of a joint task force, joint land component, or multinational headquarters should also refer to applicable joint or multinational doctrine. This includes JP 3-16, JP 3-31, and JP 3-33.

Commanders, staffs, and subordinates ensure that their decisions and actions comply with applicable United States, international, and, in some cases, host-nation laws and regulations. Commanders at all levels ensure that their Soldiers operate in accordance with the law of war and the rules of engagement. (See FM 27-10.)

ADP 5-0 implements North Atlantic Treaty Organization Standardization Agreement 2199, *Command and Control of Allied Land Forces*. ADP 5-0 uses joint terms where applicable. Selected joint and Army terms and definitions appear in both the glossary and the text. Terms for which ADP 5-0 is the proponent publication (the authority) are marked with an asterisk (*) in the glossary. When first defined in the text, terms for which ADP 5-0 is the proponent publication are boldfaced and italicized, and definitions are boldfaced. When first defining other proponent definitions in the text, the term is italicized and the number of the proponent publication follows the definition. Following uses of the term are not italicized.

ADP 5-0 applies to the Active Army, the Army National Guard/Army National Guard of the United States, and the United States Army Reserve unless otherwise stated.

The proponent of ADP 5-0 is the United States Army Combined Arms Center. The preparing agency is the Combined Arms Doctrine Directorate, United States Army Combined Arms Center. Send comments and recommendations on a DA Form 2028 (*Recommended Changes to Publications and Blank Forms*) to Commander, United States Army Combined Arms Center and Fort Leavenworth, ATTN: ATZL-MCD (ADP 5-0), 300 McPherson Avenue, Fort Leavenworth, KS 66027-2337; by e-mail to usarmy.leavenworth.mccoe.mbx.cadd-org-mailbox@mail.mil; or submit an electronic DA Form 2028.

Acknowledgements

The copyright owners listed here have granted permission to reproduce material from their works. The Source Notes lists other sources of quotations and research.

War as I Knew It by General George S. Patton. Copyright © 1947 by Beatrice Patton Walters, Ruth Patton Totten, and George Smith Totten. Copyright © renewed 1975 by MG George Patton, Ruth Patton Totten, John K. Waters, Jr., and George P. Waters. Reprinted by permission of Houghton Mifflin Company. All rights reserved.

Excerpts from *On War* by Carl von Clausewitz. Edited and translated by Peter Paret and Michael E. Howard. Copyright © 1976, renewed 2004 by Princeton University Press.

Quotes reprinted courtesy B. H. Liddell Hart, *Strategy*. Copyright © 1974 by Signet Printing.

Quote courtesy Bernard L. Montgomery, *The Memoirs of Field-Marshal Montgomery*. Copyright © 1958 by Bernard Law Montgomery. Reprinted by permission of The World Publishing Company.

Quote reprinted courtesy William Tecumseh Sherman, *Memoirs of General W. T. Sherman*. Copyright © 2000 by Penguin Books.

Quote reprinted courtesy Field-Marshall Viscount William Slim, *Defeat into Victory: Battling Japan in Burma and India, 1942–1945*. Copyright © 1956 by Viscount William Slim. Copyright © renewed 2000 by Copper Square Press.

Quote from *The Art of War* by Sun Tzu, translated by Lionel Giles. Copyright © 1910.

Quotes reprinted courtesy Stephen W. Sears' *Gettysburg*. Copyright © 2003 by Houghton Mifflin Company.

Quote reprinted courtesy Antoine Henri de Jomini, *Art of War*, translated by G.H. Mendell and W.P. Craighill. Copyright © 1862 by J.B. Lippincott & Co. Online by The Internet Archive. Available at https://archive.org/details/artwar00mendgoog.

Quote reprinted courtesy of *The American Presidency Project*. Online by Gerhard Peters and John T. Woolley. Available at https://www.presidency.ucsb.edu/node/233951.

Quote reprinted courtesy Owen Connolly, *On War and Leadership*. Copyright © 2002 by Princeton University Press.

Quotes reprinted courtesy *Dictionary of Military and Naval Quotations*, compiled by Robert Debs Heinl, Jr. Copyright © 1967 by United States Naval Institute.

Paraphrased courtesy JoAnna M. McDonald, *The Liberation of Pointe du Hoc: the 2nd U.S. Rangers at Normandy*. Copyright © 2000 by Rank and File Publications.

Quote reprinted courtesy Martin Blumenson, *The Patton Papers, vol. 2, 1940–1945*. Copyright 1974 by Martin Blumenson. Reprinted by permission of Houghton Mifflin Company. All rights reserved.

Quote from Horace Porter, *Campaigning with Grant*. Copyright © 1907 by The Century Co.

Quote courtesy Erwin Rommel, *The Rommel Papers*. Edited by B.H. Liddell-Hart. Copyright © 1953 by Harcourt, Brace, and Company.

Quote reprinted courtesy Gregory A. Daddis, *No Sure Victory: Measuring U.S. Army Effectiveness and Progress in the Vietnam War*. Copyright © 2011 by Oxford University Press.

Introduction

Military operations are human endeavors conducted in dynamic and uncertain operational environments to achieve a political purpose. Army forces, as part of a joint team, conduct unified land operations to shape operational environments, prevent conflict, consolidate gains, and contribute to winning the Nation's wars. During periods of competition or armed conflict, command and control—the exercise of authority and direction by a properly designated commander—is fundamental to all operations. Based on the Army's vision of war and nature of operations, mission command is the Army's approach for exercising command and control. The mission command approach empowers subordinate decision making and emphasizes decentralized execution appropriate to the situation.

The Army's framework for organizing and putting command and control into action is the operations process. The operations process consists of the major command and control activities performed during operations (planning, preparing, executing, and continuously assessing). Commanders, supported by their staffs, employ the operations process to understand, visualize, and describe their operational environments, end state, and operational approach; make and articulate decisions; and direct, lead, and assess military operations as shown in introduction figure-1 on page vi.

The Army continuously prepares for large-scale ground combat while simultaneously shaping the security environment around the world. ADP 5-0 provides doctrine for how Army forces conduct the operations process across the range of military operations. It describes a mission command approach to planning, preparing, executing, and assessing operations. This revised ADP 5-0—

- Combines the 2012 editions of ADP 5-0 and ADRP 5-0 into one publication.
- Incorporates updated tactics on Army operations to include an emphasis on large-scale combat operations described in the 2017 edition of FM 3-0.
- Incorporates updated fundamentals of mission command to include the reintroduction of *command and control* to Army doctrine described in the 2019 edition of ADP 6-0.
- Incorporates updated doctrine on assessment described in the 2017 edition of JP 5-0.
- Removes the detailed discussion of Army design methodology (now found in ATP 5-0.1).
- Removes the discussion of continuing activities as they are similar to the responsibilities of units assigned an area of operations.

ADP 5-0 contains five chapters:

Chapter 1 sets the context for conducting the operations process by describing the nature of operations, unified land operations, and mission command. Next, it defines and describes the operations process. A discussion of the principles of the operations process follows. The chapter concludes with discussions of the integrating processes and battle rhythm.

Chapter 2 defines planning and describes the functions of planning and plans. It discusses planning at the levels of warfare, operational art, integrated planning, and key components of a plan. The chapter concludes with guides for effective planning and planning pitfalls to avoid.

Chapter 3 addresses the fundamentals of preparation to include its definition and functions. It offers guides for effective preparation and addresses specific preparation activities commonly performed within the headquarters and across the force to improve the unit's ability to execute operations.

Chapter 4 defines, describes, and offers guides for effective execution. It describes the role of the commander and the role of the staff during execution followed by a discussion of the major activities of execution. It concludes with a discussion of the rapid decision-making and synchronization process.

Chapter 5 defines and describes assessment. It discusses an assessment model and offers guides for effective assessment.

Introduction

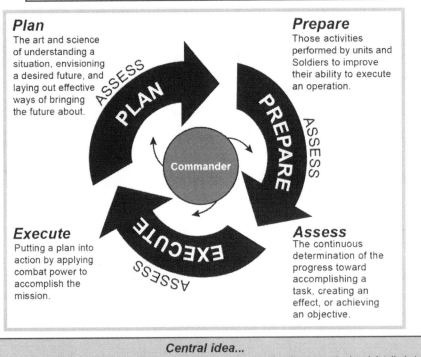

Introduction figure-1. Operations process logic chart

Introduction table-1 lists terms that have been added, rescinded, or modified for which ADP 5-0 is the proponent publication for the term. The glossary contains acronyms and defined terms.

Introduction table-1. New, modified, and removed Army terms

Term	Reasoning
Army design methodology	Modifies for clarity.
civil considerations	ADP 6-0 becomes proponent.
collaborative planning	ADP 5-0 becomes proponent and modifies the definition.
commander's visualization	ADP 6-0 becomes proponent.
concept of operations	ADP 5-0 becomes proponent.
confirmation brief	ADP 5-0 becomes proponent and modifies the definition.
decision support matrix	ADP 5-0 becomes proponent.
direct support	FM 3-0 becomes proponent for the Army term.
essential element of friendly information	ADP 6-0 becomes proponent.
evaluating	ADP 5-0 becomes proponent and modifies the definition.
execution	Modifies for clarity.
execution matrix	ADP 5-0 becomes proponent and modifies the definition.
general support-reinforcing	FM 3-0 becomes proponent.
indicator	Army definition is no longer used. Adopts joint definition.
key tasks	ADP 6-0 becomes proponent.
military decision-making process	Modifies term for grammar.
monitoring	ADP 5-0 becomes proponent.
nested concepts	ADP 5-0 becomes proponent.
operations process	Modifies for clarity.
parallel planning	ADP 5-0 becomes proponent and modifies the definition.
planning	Modifies for clarity.
planning horizon	ADP 5-0 becomes proponent.
priority of support	ADP 5-0 becomes proponent.
rehearsal	ADP 5-0 becomes proponent and modifies the definition.
reinforcing	FM 3-0 becomes proponent.
situational understanding	ADP 6-0 becomes proponent.
task organization	ADP 5-0 becomes proponent.
terrain management	ADP 3-90 becomes proponent.

This page intentionally left blank.

Chapter 1
Fundamentals of the Operations Process

The best is the enemy of good. By this I mean that a good plan violently executed now is better than a perfect plan next week.

General George S. Patton, Jr.

This chapter sets the contexts for conducting the operations process by describing the nature of operations, unified land operations, and mission command. Next, it defines and describes the operations process. A discussion of the principles of the operations process follows. The chapter concludes with discussions of the integrating processes and battle rhythm.

THE NATURE OF OPERATIONS

1-1. Understanding the doctrine on mission command and the operations process requires an appreciation of the nature of operations and the Army's vision of war. It is upon this appreciation that mission command—an approach to the exercise of command and control—is built. The principles of mission command guide commanders and staffs in planning, preparing, executing, and assessing operations.

1-2. Military operations fall along a competition continuum that spans cooperation to war. Between these extremes, societies maintain relationships. These relationships include economic competition, political or ideological tension, and at times armed conflict. Violent power struggles in failed states, along with the emergence of major regional powers like Russia, China, Iran, and North Korea seeking to gain strategic positions of advantage, present challenges to the joint force. Army forces must be prepared to meet these challenges across the range of military operations during periods of competition and war.

1-3. The range of military operations is a fundamental construct that helps relate military activities and operations in scope and purpose within a backdrop of the competition continuum. The potential range of military operations extends from military engagement, security cooperation, and deterrence in periods of competition to large-scale combat operations in periods of war. Whether fighting terrorists as part of a limited contingency operation or defeating a peer threat in large-scale combat, the nature of operations is constant. Military operations are—
- Human endeavors.
- Conducted in dynamic and uncertain environments.
- Designed to achieve a political purpose.

HUMAN ENDEAVORS

1-4. Military operations are human endeavors—a contest of wills characterized by violence and continuous adaptation among all participants. Fundamentally, all war is about changing human behavior. During operations, Army forces face thinking and adaptive enemies, differing agendas of various actors, and changing perceptions of civilians in an operational area. The enemy is not an inanimate object to be acted upon but an independent and active force with its own objectives. As friendly forces try to impose their will on the enemy, the enemy resists and seeks to impose its will on friendly forces. A similar dynamic occurs among civilian groups whose own desires influence and are influenced by military operations. Appreciating these relationships among opposing human wills is essential to understanding the fundamental nature of operations.

Dynamic and Uncertain

Everything in war is very simple, but the simplest thing is difficult. The difficulties accumulate and end by producing a kind of friction that is inconceivable unless one has experienced war.

Carl von Clausewitz

1-5. War is inherently dynamic and uncertain. The complexity of friendly and enemy organizations, unique combinations of terrain and weather, and the dynamic interaction among all participants create uncertainty. Chance and friction increase the potential for chaos and uncertainty during operations. Chance pertains to unexpected events or changes beyond the control of friendly forces, while friction describes obstacles that make executing even simple tasks difficult. Both are always present for all sides during combat.

1-6. The scale, scope, tempo, and lethality of large-scale ground combat exacerbates the dynamic and uncertain nature of war, making precise cause-and-effect determinations difficult or delayed. The unintended effects of operations often cannot be anticipated and may not be readily apparent during execution. Disorder is an inherent characteristic of war. This demands an approach to the conduct of operations that does not attempt to impose perfect order on operations but rather makes allowances for friction and uncertainty.

Achieve Political Purpose

Thus any study of the problem ought to begin and end with the question of policy.

Sir Basil Henry Liddell Hart

1-7. All U.S. military operations share a common fundamental purpose—to achieve or contribute to national objectives. Objective—to direct every military operation toward a clearly defined, decisive, and attainable goal—is a principle of war. This principle reinforces the proper relationship between military operations and policy. Military operations must always be subordinate to policy and serve as a way to a political end.

1-8. In large-scale combat, the purpose of operations may be to destroy the enemy's capabilities and will to fight. The purpose of operations short of large-scale combat may be more nuanced and broad, and subsequently, may require support to multiple objectives. These operations frequently involve setting conditions that improve positions of relative advantage compared to that of a specific adversary and that contribute to achieving strategic aims in an operational area. In either case, all operations are designed to achieve the political purpose set by national authorities.

UNIFIED LAND OPERATIONS

1-9. The Army's operational concept—the central idea that guides the conduct of Army operations—is unified land operations. *Unified land operations* is the simultaneous execution of offense, defense, stability, and defense support of civil authorities across multiple domains to shape operational environments, prevent conflict, prevail in large-scale ground combat, and consolidate gains as part of unified action (ADP 3-0). Army forces do this with combined arms formations possessing the mobility, firepower, protection, and sustainment to defeat an enemy and establish control of areas, resources, and populations. Army forces depend on the capabilities of the other Services as the joint force depends on Army capabilities across multiple domains. The goal of unified land operations is to achieve the joint force commander's end state by applying land power as part of unified action. During the conduct of unified land operations, Army forces support the joint force through four strategic roles:

- Shape operational environments (OEs).
- Prevent conflict.
- Prevail in large-scale ground combat.
- Consolidate gains.

1-10. Army forces assist in shaping an OE by providing trained and ready forces to geographic combatant commanders (GCCs) in support of their campaign plan. Shaping activities include security cooperation, military engagement, and forward presence to promote U.S. interests and assure allies. Army operations to prevent are designed to deter undesirable actions of an adversary through positioning of friendly capabilities and demonstrating the will to use them. Army forces may have a significant role in the execution of flexible deterrent options or flexible response options. Additionally, Army prevent activities may include mobilization, force tailoring, and other pre-deployment activities; initial deployment into a theater of operations; and development of intelligence, communications, sustainment, and protection infrastructure to support the joint force commander. During large-scale combat operations, Army forces focus on the defeat of enemy ground forces. Army forces close with and destroy enemy forces, exploit success, and break their opponent's will to resist. While Army forces consolidate gains throughout an operation, consolidating gains become the focus of operations after large-scale combat operations have concluded. (See ADP 3-0 for a detailed discussion of unified land operations.)

MISSION COMMAND

Diverse are the situations under which an officer has to act on the basis of his own view of the situation. It would be wrong if he had to wait for orders at times when no orders can be given. But most productive are his actions when he acts within the framework of his senior commander's intent.

Field Marshal Helmuth von Moltke

1-11. *Command and control* is the exercise of authority and direction by a properly designated commander over assigned and attached forces in the accomplishment of the mission (JP 1). Command and control is fundamental to all operations. By itself, however, command and control will not secure an objective, destroy an enemy target, or deliver supplies. Yet none of these activities could be coordinated towards a common objective, or synchronized to achieve maximum effect, without effective command and control. It is through command and control that the countless activities a military force must perform gain purpose and direction. The goal of command and control is effective mission accomplishment.

1-12. *Mission command* is the Army's approach to command and control that empowers subordinate decision making and decentralized execution appropriate to the situation (ADP 6-0). Mission command is based on the Army's view that war is inherently chaotic and uncertain. No plan can account for every possibility and most plans must change rapidly during execution if they are to succeed. No single person is ever well-enough informed to make every important decision, nor can a single person manage the number of decisions that need to be made during combat. As such, mission command empowers subordinate leaders to make decisions and act within the commander's intent to exploit opportunities and counter threats.

1-13. Mission command requires an environment of trust and shared understanding among commanders, staffs, and subordinates. It requires building effective teams and a command climate in which commanders encourage subordinates to accept risk and exercise initiative to seize opportunities and counter threats within the commander's intent. Through mission orders, commanders focus leaders on the purpose of the operation rather than on the details of how to perform assigned tasks. Doing this minimizes detailed control and allows subordinates the greatest possible freedom of action to accomplish tasks. Finally, when delegating authority to subordinates, commanders set the necessary conditions for success by allocating appropriate resources to subordinates based on assigned tasks.

1-14. Because uncertainty is pervasive during operations, success is often determined by a leader's ability to outthink an opponent and to execute tasks more quickly than an opponent can react. The side that anticipates better, thinks more clearly, decides and acts more quickly, and is comfortable operating with uncertainty stands the greatest chance to seize, retain, and exploit the initiative over an opponent. Leaders make decisions, develop plans, and direct actions with the information they have at the time. Commanders seek to counter the uncertainty of operations by empowering subordinates to quickly adapt to changing circumstances within their intent. Mission command decentralizes decision-making authority and grants subordinates significant freedom of action. The principles of mission command are—

- Competence.
- Shared understanding.
- Mutual trust.
- Mission orders.

Chapter 1

- Commander's intent.
- Disciplined initiative.
- Risk acceptance.

(See ADP 6-0 for a detailed discussion of mission command.)

THE FRAMEWORK OF THE OPERATIONS PROCESS

1-15. The Army's framework for organizing and putting command and control into action is the *operations process*—**the major command and control activities performed during operations: planning, preparing, executing, and continuously assessing the operation.** Commanders use the operations process to drive the conceptual and detailed planning necessary to understand their OE; visualize and describe the operation's end state and operational approach; make and articulate decisions; and direct, lead, and assess operations as shown in figure 1-1.

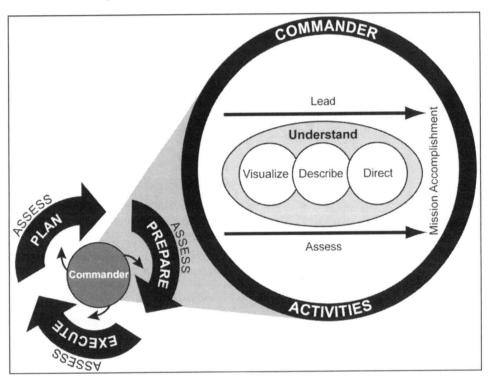

Figure 1-1. The operations process

1-16. Commanders, staffs, and subordinate headquarters employ the operations process to organize efforts, integrate the warfighting functions across multiple domains, and synchronize forces to accomplish missions. This includes integrating numerous processes and activities such as information collection and targeting within the headquarters and with higher, subordinate, supporting, and supported units. The unit's battle rhythm (see the discussion beginning in paragraph 1-82) helps to integrate and synchronize the various processes and activities that occur within the operations process.

1-17. A goal of the operations process is to make timely and effective decisions and to act faster than the enemy. A tempo advantageous to friendly forces can place the enemy under the pressures of uncertainty and time. Throughout the operations process, making and communicating decisions faster than the enemy can react produces a tempo with which the enemy cannot compete. These decisions include assigning tasks; prioritizing, allocating, and organizing forces and resources; and selecting the critical times and places to act. Decision making during execution includes knowing how and when to adjust previous decisions. The speed and accuracy of a commander's actions to address a changing situation is a key contributor to agility.

> ### Agility: Rapidly Turning the Third Army to Bastogne
>
> The summer and fall of 1944 saw significant gains by Allied armies in Western Europe. Since D-Day, the German army in the west had been falling steadily back toward the Rhine River. Allied commanders were confident that the war would be over in a matter of months. On 16 December 1944, the Wehrmacht launched three German armies in OPERATION WATCH ON THE RHINE against the U.S. First Army. This enemy counteroffensive caught the Allies off guard, causing significant operational changes in order to stabilize the penetration to the center of the Allied front.
>
> Lieutenant General George Patton was leading the U.S. Third Army in offensive operations south of the First Army when the German counteroffensive began. Initially unconcerned with the reports of a German attack, Patton began to take note of German gains as the situation developed and started preparing to respond to the threat. "On the eighteenth, [General Omar] Bradley called me to come to Luxembourg ... for a conference. ...he showed me that the German penetration was much greater than we had thought and asked what I could do." Patton told Bradley he could have three divisions moving north within twenty-four hours. After receiving his assessment, Bradley directed Patton to attend a meeting the next day with General Dwight Eisenhower to discuss options.
>
> Returning to his headquarters, Patton met with his staff to prepare for meeting Eisenhower. Patton started the meeting by stating "plans had been changed, and, while we were all accustomed to rapid movement, we would now have to prove that we could operate even faster." He then directed his staff to develop three options for attacking north as soon as possible.
>
> Upon meeting Eisenhower and the collected Allied staff, Patton stated he could attack the German forces with a three-division front by the 22nd of December. Once it was decided to allow Patton to attack, Patton called his headquarters to start movement of the three divisions. He used the preplanned code phrase assigned to one of the options being planned back at his headquarters. The forethought of Patton and his staff allowed him to rapidly change battle plans enabling elements of the Third Army to attack north, relieve elements of the First Army, and cut off the German army in what is known as the Battle of the Bulge.

1-18. Both the commander and staff have important roles within the operations process. The commander's role is to drive the operations process through the activities of understanding, visualizing, describing, directing, leading, and assessing operations as described in paragraphs 1-31 through 1-49. The staff's role is to assist commanders with understanding situations, making and implementing decisions, controlling operations, and assessing progress. In addition, the staff assists subordinate units (commanders and staffs), and keeps units and organizations outside the headquarters informed throughout the conduct of operations. (See FM 6-0 for a detailed discussion of the duties and responsibilities of the staff.)

1-19. The Army Ethic guides commanders, leaders, and staffs throughout the operations process. The Army Ethic is the evolving set of laws, values, and beliefs, embedded within the Army culture of trust that motivates and guides the conduct of Army professionals bound together in common moral purpose. The Army demands its members to make ethical, effective, and efficient decisions and to act according to the moral principles of its ethic. The Uniform Code of Military Justice, Army regulations, the law of war, rules of engagement, and the Code of Conduct set the minimum standards for ethical conduct. (See ADP 6-22 for a discussion of the Army Ethic.)

Chapter 1

ACTIVITIES OF THE OPERATIONS PROCESS

It is a mistake to think that once an order is given there is nothing more to be done; you have got to see that it is carried out in the spirit which you intended.

Field Marshal Bernard L. Montgomery

1-20. The activities of the operations process are not discrete; they overlap and recur as circumstances demand. While planning may start an iteration of the operations process, planning does not stop with the production of an order. After the completion of the initial order, the commander and staff continuously revise the plan based on changing circumstances. Preparation for a specific mission begins early in planning and continues for some subordinate units during execution. Execution puts a plan into action and involves adjusting the plan based on changes in the situation and the assessment of progress. Assessing is continuous and influences the other three activities.

Planning

1-21. Planning is the art and science of understanding a situation, envisioning a desired future, and laying out effective ways of bringing that future about. Planning is both conceptual and detailed. Conceptual planning includes developing an understanding of an OE, framing the problem, defining a desired end state, and developing an operational approach to achieve the desired end state. Conceptual planning generally corresponds to the art of operations and is commander led. Detailed planning translates the operational approach into a complete and practical plan. Detailed planning generally corresponds to the science of operations and encompasses the specifics of implementation. Detailed planning works out the scheduling, coordination, or technical issues involved with moving, sustaining, administering, and directing forces. (See chapter 2 for the fundamentals of planning.)

Preparation

1-22. Preparation consists of activities that units and Soldiers perform to improve their abilities to execute an operation. Preparation creates conditions that improve friendly forces' opportunities for success. Activities of preparation help develop a shared understanding of the situation and requirements for execution. These activities—such as backbriefs, rehearsals, training, and inspections—help units, staffs, and Soldiers better understand their roles in upcoming operations, gain proficiency on complicated tasks, and ensure their equipment and weapons function properly. (See chapter 3 for the fundamentals of preparation.)

Execution

1-23. Planning and preparation enable effective execution. Execution is putting a plan into action while using situational understanding to assess progress and adjust operations as the situation changes. Execution focuses on concerted action to seize and retain the initiative, build and maintain momentum, and exploit success. (See chapter 4 for the fundamentals of execution.)

Assessment

1-24. Assessment precedes and guides the other activities of the operations process and concludes each operation or phase of an operation. The focus of assessment differs during planning, preparation, and execution. During planning, assessment focuses on gathering information to understand the current situation and developing an assessment plan. During preparation, assessment focuses on monitoring changes in the situation and on evaluating the progress of readiness to execute the operation. Assessment during execution involves a deliberate comparison of forecasted outcomes to actual events, using criterion to judge progress toward success. Assessment during execution helps commanders adjust plans based on changes in the situation. (See chapter 5 for the fundamentals of assessment.)

CHANGING CHARACTER OF THE OPERATIONS PROCESS

1-25. The situation and type of operations affects the character of the operations process. For example, planning horizons (a point in time commanders use to focus the organization's planning efforts) and decision cycles are generally shorter in large-scale combat operations than in a counterinsurgency operation. The accelerated tempo, hyperactive chaos, and lethality of large-scale combat operations often require rapid decision making and synchronization to exploit opportunities and reduce risk. During large-scale ground combat, command posts displace often, communications are degraded, and troops receive limited precise information about the enemy. These conditions influence the operations process. Streamlining staff processes and the unit's battle rhythm to those related to the defeat of the enemy is essential. Counterinsurgency operations are generally more methodical and deliberate. For example, they often have consolidated and stationary headquarters, longer planning horizons, and more time available for information gathering and analysis to inform decision making.

1-26. The character of the operations process also varies depending on echelon. Higher echelons generally have longer planning horizons and often have to make decisions concerning operations well in advance of execution. For example, a theater army commander plans for and requests forces months before their anticipated employment in follow-on phases of an operation. A division commander may decide to change task-organization days in advance of an operation to allow time for units to reposition and integrate into their new formation. On the other hand, changing the direction of an attack for a combined arms battalion requires limited planning and can be executed in hours or minutes.

1-27. Depending on the echelon, all activities of the operations process can occur simultaneously within a headquarters. Divisions and corps headquarters are staffed with a plans cell, a future operations cell, and a current operations integration cell. These headquarters can plan, prepare, execute, and assess operations simultaneously, cycling through multiple iterations of the operations process. Companies however, tend to move sequentially through the activities of the operations process because they lack a staff. A company commander receives the mission and conducts troop leading procedures (TLP). After developing the plan, the company commander conducts rehearsals and supervises preparation prior to execution. The company then executes its mission while continuously assessing. Following execution, the company consolidates and reorganizes in preparation of a new mission starting a new cycle of the operations process.

MULTINATIONAL OPERATIONS AND THE OPERATIONS PROCESS

1-28. *Multinational operations* is a collective term to describe military actions conducted by forces of two or more nations, usually undertaken within the structure of a coalition or alliance (JP 3-16). Multinational operations are driven by common agreement among the participating alliance or coalition partners. While each nation has its own interests and often participates within the limitations of national caveats, all nations bring value to an operation. Each nation's force has unique capabilities, and each usually contributes to the operation's legitimacy in terms of international or local acceptability. Army forces should anticipate that most operations will be multinational operations and plan accordingly.

1-29. Multinational operations present challenges and demands throughout the operations process. These include cultural and language issues, interoperability challenges, national caveats on the use of respective forces, the sharing of information and intelligence, and rules of engagement. Establishing standard operating procedures (SOPs) and liaison with multinational partners is critical to effective command and control. When conducting the operations process within a multinational training or operational setting, Army commanders should be familiar with and employ multinational doctrine and standards ratified by the U.S. For example, Allied Tactical Publication 3.2.2, *Command and Control of Land Forces*, applies to Army forces during the conduct of North Atlantic Treaty Organization (known as NATO) operations. (See FM 3-16 for a detailed discussion on multinational operations.)

PRINCIPLES OF THE OPERATIONS PROCESS

1-30. The operations process, while simple in concept, is dynamic in execution. Commanders must organize and train their staffs and subordinates as an integrated team to simultaneously plan, prepare, execute, and assess operations. In addition to the principles of mission command, commanders and staffs consider the following principles for the effective employment of the operations process:
- Drive the operations process.
- Build and maintain situational understanding.
- Apply critical and creative thinking.

Chapter 1

DRIVE THE OPERATIONS PROCESS

1-31. Commanders are the most important participants in the operations process. While staffs perform essential functions that amplify the effectiveness of operations, commanders drive the operations process through understanding, visualizing, describing, directing, leading, and assessing operations. Accurate and timely running estimates maintained by the staff, assist commanders in understanding situations and making decisions. See figure 1-2.

Understand

1-32. Understanding an OE and associated problems is fundamental to establishing a situation's context and visualizing operations. An *operational environment* is a composite of the conditions, circumstances, and influences that affect the employment of capabilities and bear on the decisions of the commander (JP 3-0). An OE encompasses the air, land, maritime, space, and cyberspace domains; the information environment; the electromagnetic spectrum; and other factors. Included within these areas are the enemy, friendly, and neutral actors who are relevant to a specific operation.

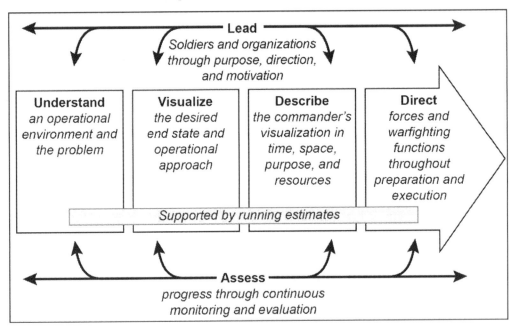

Figure 1-2. The commander's role in the operations process

1-33. Commanders collaborate with their staffs, other commanders, and unified action partners to build a shared understanding of their OEs and associated problems. Planning, intelligence preparation of the battlefield (IPB), and running estimates help commanders develop an initial understanding of their OEs. During execution, commanders direct reconnaissance and develop the situation through action to improve their understanding. Commanders circulate within the area of operations (AO) as often as possible, collaborating with subordinate commanders and speaking with Soldiers. Ideally, true understanding should be the basis for decisions. However, commanders realize that uncertainty and time often preclude their achieving complete understanding before deciding and acting.

Visualize

1-34. As commanders build understanding about their OEs, they start to visualize solutions to solve the problems they identify. Collectively, this is known as *commander's visualization*—the mental process of developing situational understanding, determining a desired end state, and envisioning an operational approach by which the force will achieve that end state (ADP 6-0).

1-35. In building their visualization, commanders first seek to understand those conditions that represent the current situation. Next, commanders envision a set of desired future conditions that represents the operation's end state. Commanders complete their visualization by conceptualizing an *operational approach*—a broad description of the mission, operational concepts, tasks, and actions required to accomplish the mission (JP 5-0). Figure 1-3 depicts activities associated with developing the commander's visualization.

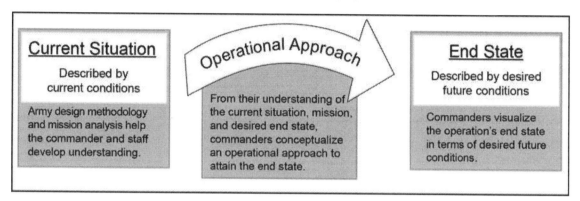

Figure 1-3. Commander's visualization

1-36. Part of developing an operational approach includes visualizing an initial operational framework. The operational framework provides an organizing construct for how the commander intends to organize the AO geographically (deep, close, support, and consolidation areas), by purpose (decisive, shaping, and sustaining operations), and by effort (main and supporting). When establishing their operational framework, commanders consider the physical, temporal, virtual, and cognitive factors that impact on their AOs. Collectively, these considerations allow commanders and staffs to better account for the multi-domain capabilities of friendly and threat forces. (See ADP 3-0 for a detailed discussion of the operational framework.)

Describe

1-37. Commanders describe their visualization to their staffs and subordinate commanders to facilitate shared understanding and purpose throughout the force. During planning, commanders ensure subordinates understand their visualization well enough to begin course of action (COA) development. During execution, commanders describe modifications to their visualization in updated planning guidance and directives resulting in fragmentary orders (FRAGORDs) that adjust the original operation order (OPORD). Commanders describe their visualization in doctrinal terms, refining and clarifying it, as circumstances require. Commanders describe their visualization in terms of—

- Commander's intent.
- Planning guidance.
- Commander's critical information requirements (CCIRs).
- Essential elements of friendly information.

Commander's Intent

> *I suppose dozens of operation orders have gone out in my name, but I never, throughout the war, actually wrote one myself. I always had someone who could do that better than I could. One part of the order I did, however, draft myself—the intention.*
>
> Field-Marshall Viscount William Slim

1-38. The *commander's intent* is a clear and concise expression of the purpose of the operation and the desired military end state that supports mission command, provides focus to the staff, and helps subordinate and supporting commanders act to achieve the commander's desired results without further orders, even when the operation does not unfold as planned (JP 3-0). During planning, the initial commander's intent guides COA development. In execution, the commander's intent guides initiative as subordinates make decisions and take action when unforeseen opportunities arise or when countering threats. Commanders

Chapter 1

develop their intent statement personally. It must be easy to remember and clearly understood by commanders and staffs two echelons lower in the chain of command. The more concise the commander's intent, the easier it is to understand and recall.

Commander's Planning Guidance

1-39. Commanders provide planning guidance to the staff based upon their visualization of the operation. Planning guidance conveys the essence of the commander's visualization, including a description of the operational approach. Effective planning guidance reflects how the commander sees the operation unfolding. The commander's planning guidance broadly describes when, where, and how the commander intends to employ combat power to accomplish the mission within the higher commander's intent. Broad and general guidance gives the staff and subordinate leaders maximum latitude; it lets proficient staffs develop flexible and effective options. Commanders modify planning guidance based on staff and subordinate input and changing conditions during different stages of planning and throughout the operations process. (See FM 6-0 for sample planning guidance by warfighting function.)

Commander's Critical Information Requirements

1-40. A *commander's critical information requirement* is an information requirement identified by the commander as being critical to facilitating timely decision making (JP 3-0). Commanders decide to designate an information requirement as a CCIR based on likely decisions during the conduct of an operation. A CCIR may support one or more decision points. During planning, staffs recommend information requirements for commanders to designate as CCIRs. During preparation and execution, they recommend changes to CCIRs based on their assessments of the operation.

1-41. Always promulgated by a plan or order, commanders limit the number of CCIRs to focus their staff and subordinate unit information collection and assessment efforts. The fewer the CCIRs, the easier it is for staffs to remember, recognize, and act on each one. As such, the rapid reporting of CCIRs to the commander is essential to adjusting operations. CCIR falls into one of two categories: priority intelligence requirements (known as PIRs) and friendly force information requirements (known as FFIRs).

1-42. A *priority intelligence requirement* is an intelligence requirement that the commander and staff need to understand the threat and other aspects of the operational environment (JP 2-01). Priority intelligence requirements identify the information about the enemy and other aspects of an OE that the commander considers most important. Intelligence about civil considerations may be as critical as intelligence about the enemy. In coordination with the staff, the intelligence officer manages priority intelligence requirements for the commander as part of the intelligence process.

1-43. A *friendly force information requirement* is information the commander and staff need to understand the status of friendly force and supporting capabilities (JP 3-0). Friendly force information requirements identify the information about the mission, troops and support available, and time available for friendly forces that the commander considers most important. In coordination with the staff, the operations officer manages friendly force information requirements for the commander.

Essential Elements of Friendly Information

1-44. Commanders also describe information they want protected as essential elements of friendly information. An *essential element of friendly information* is a critical aspect of a friendly operation that, if known by a threat would subsequently compromise, lead to failure, or limit success of the operation and therefore should be protected from enemy detection (ADP 6-0). Although essential elements of friendly information (known as EEFIs) are not CCIRs, they have the same priority. Essential elements of friendly information establish elements of information to protect rather than elements to collect. Their identification is the first step in the operations security process and central to the protection of information.

Direct

1-45. To direct is implicit in command. Commanders direct action to achieve results and lead forces to mission accomplishment. Commanders make decisions and direct action based on their situational understanding maintained by continuous assessment. Throughout the operations process, commanders direct forces by—

- Approving plans and orders.
- Establishing command and support relationships.
- Assigning and adjusting tasks, control measures, and task organization.
- Positioning units to maximize combat power.
- Positioning key leaders at critical places and times to ensure supervision.
- Allocating resources to exploit opportunities and counter threats.
- Committing the reserve.

Lead

Example whether it be good or bad has a powerful influence.

General George Washington

1-46. *Leadership* is the activity of influencing people by providing purpose, direction, and motivation to accomplish the mission and improve the organization (ADP 6-22). Leadership inspires Soldiers to accomplish things that they otherwise might not. Throughout the operations process, commanders make decisions and provide the purpose and motivation to follow through with the COA they chose. They must also possess the wisdom to know when to modify a COA when situations change. (See ADP 6-22 for a detailed discussion of leadership to include attributes of effective leaders.)

1-47. Commanders lead by example through command presence. Command presence is creating a favorable impression in demeanor, appearance, and professional and personal conduct. Commanders use their presence to gather and communicate information and knowledge as well as to assess operations. Establishing a command presence makes the commander's knowledge and experience available to subordinates. It allows commanders to evaluate and provide direct feedback on their subordinates' performance.

1-48. Command occurs at the location of the commander. Where the commander locates within the AO is an important consideration for effective mission command. No standard pattern or simple prescription exists for the proper location of a commander on the battlefield; different commanders lead differently. Commanders balance their time among the command post and staff, subordinate commanders, forces, and other organizations to make the greatest contribution to success. (See ADP 6-0 for discussions of command presence and location of the commander during operations.)

Assess

1-49. Assessment involves deliberately comparing intended forecasted outcomes with actual events to determine the overall effectiveness of force employment. Assessment helps the commander determine progress toward attaining the desired end state, achieving objectives, and completing tasks. Commanders incorporate assessments by the staff, subordinate commanders, and unified action partners into their personal assessment of the situation. Based on their assessment, commanders adjust their visualization and modify plans and orders to adapt the force to changing circumstances. (See chapter 5 for a detailed discussion on assessment.)

BUILD AND MAINTAIN SITUATIONAL UNDERSTANDING

1-50. Success in operations demands timely and effective decisions based on applying judgment to available information and knowledge. As such, commanders and staffs seek to build and maintain situational understanding throughout the operations process. *Situational understanding* is the product of applying analysis and judgment to relevant information to determine the relationships among the operational and mission variables (ADP 6-0). Commanders and staffs continually strive to maintain their situational understanding and work through periods of reduced understanding as a situation evolves. Effective commanders accept that uncertainty can never be eliminated and train their staffs and subordinates to function in uncertain environments.

Chapter 1

1-51. As commanders build their situational understanding, they share their understanding across the forces and with unified action partners. Creating shared understanding is a principle of mission command and requires communication and information sharing from higher to lower and lower to higher. Higher headquarters ensure subordinates understand the larger situation to include the operation's end state, purpose, and objectives. Staffs from lower echelons share their understanding of their particular situation and provide feedback to the higher headquarters on the operation's progress. Communication and information sharing with adjacent units and unified action partners is also multi-directional. Several tools assist leaders in building situational understanding and creating a shared understanding across the force to include—

- Operational and mission variables.
- Running estimates.
- Intelligence.
- Collaboration.
- Liaison.

Operational and Mission Variables

1-52. Operational and mission variables are categories of relevant information commanders and staffs use to help build their situational understanding. Commanders and staffs use the eight interrelated operational variables—political, military, economic, social, information, infrastructure, physical environment, and time (known as PMESII-PT)—to help understand an OE. Operational variables are those aspects of an OE, both military and nonmilitary, that may differ from one operational area to another and affect operations.

1-53. Upon receipt of a mission, commanders and staffs filter information categorized by the operational variables into relevant information with respect to the mission. They use the mission variables, in combination with the operational variables, to refine their understanding of their situation and to visualize, describe, and direct operations. The mission variables are mission, enemy, terrain and weather, troops and support available, time available, and civil considerations (known as METT-TC). Commanders and staffs view all the mission variables in terms of their impact on mission accomplishment. (See FM 6-0 for a detailed description of the operational and mission variables.)

Running Estimates

1-54. A *running estimate* **is the continuous assessment of the current situation used to determine if the current operation is proceeding according to the commander's intent and if planned future operations are supportable.** Running estimates assist commanders and staffs with understanding situations, assessing progress, and making decisions throughout an operation. Effective plans and successful execution hinge on current and accurate running estimates.

1-55. Each staff section maintains a running estimate within its specified area of expertise (for example, intelligence, fires, logistics, and personnel). When building and maintaining a running estimate, staff sections monitor current operations and continuously consider the following in context of the operations:

- Facts.
- Assumptions.
- Friendly status including location, activity, and combat power of subordinate units from two levels down.
- Enemy status including composition, disposition, and strength.
- Civil considerations.
- Conclusions and recommendations.

1-56. Running estimates cover essential facts and assumptions, including a summary of the current situation. Running estimates always include recommendations for anticipated decisions. During planning, commanders use these recommendations to select valid (feasible, acceptable, suitable, distinguishable, and complete) COAs for further analysis. During preparation and execution, commanders use recommendations from running estimates to inform their decision making.

1-57. While staffs maintain formal running estimates, the commander's estimate is a mental process directly tied to the commander's visualization. Commanders integrate personal knowledge of the situation, analysis of the mission variables, assessments by subordinate commanders and other organizations, and relevant details gained from running estimates.

1-58. Because a commander may need a running estimate at any time, staffs must develop, update, and continuously revise running estimates while in garrison and during operations. At a minimum, staffs maintain a running estimate on friendly capabilities while in garrison or when not actively engaged in operations. Commanders and staff elements immediately begin updating their running estimates upon receipt of a mission. They continue to build and maintain their running estimates throughout the operations process in planning, preparation, execution, and assessment.

Intelligence

If you know the enemy and know yourself, you need not fear the result of a hundred battles.

Sun Tzu

1-59. Intelligence supports the commander and staff in building and maintaining situational understanding during all activities of the operations process. Information and intelligence are essential for developing an understanding of the threat, terrain and weather, and civil considerations. Intelligence helps commanders understand and visualize their OEs and options available to the enemy and the friendly force.

1-60. The intelligence process describes how the intelligence warfighting function facilitates situational understanding and supports decision making. This process provides a common framework for Army professionals to guide their thoughts, discussions, plans, and assessments. Effective execution of the intelligence process depends on commander and staff involvement and effective information collection. Commanders drive the intelligence process by issuing planning guidance, establishing priorities, identifying decision points, and designating their CCIRs. The intelligence process generates information, products, and knowledge about an OE during planning, preparation, execution, and assessment. It also integrates intelligence into targeting, information operations, and risk management. (See ADP 2-0 for a detailed discussion of the intelligence process.)

Collaboration

1-61. Commanders and staffs actively build and maintain shared understanding within the force and with unified action partners by continually collaborating throughout the operations process. Collaboration is more than coordination. It is multiple people and organizations working together towards a common goal by sharing knowledge and building consensus. It requires dialogue that involves a candid exchange of ideas or opinions among participants and encourages frank discussions in areas of disagreement. Throughout the operations process, commanders, subordinate commanders, staffs, and unified action partners collaborate, sharing and questioning information, perceptions, and ideas to understand situations and make decisions.

1-62. Through collaboration, the commander creates a learning environment by allowing participants to think critically and creatively and share their ideas, opinions, and recommendations without fear of reproach. Effective collaboration requires candor and a free, yet mutually respectful, exchange of ideas. Participants must feel free to make viewpoints based on their expertise, experience, and insight. This includes sharing ideas that contradict the opinions held by those of higher rank. Successful commanders listen to novel ideas and counterarguments. Effective collaboration is not possible unless the commander enables it.

Chapter 1

> ## Collaboration: Meade's Council of War
>
> In June 1863, General Robert E. Lee prepared the Army of Northern Virginia for a second invasion of the North. Moving through the Shenandoah Valley and north toward Harrisburg, Lee's Army made contact with the Army of the Potomac near the town of Gettysburg on July 1, 1863. Day one of the battle saw initial Confederate success. By the afternoon of day two, Major General George Meade (who had just recently assumed command of the Army of the Potomac) had moved the bulk of his force into defensive positions on the high ground south of the city. The battlefield was set.
>
> Late in the afternoon of July 2, Lee launched heavy assaults on both the Union's left and right flanks. Fierce fighting raged at Little Round Top, the Wheatfield, Devil's Den, Culp's Hill, and Cemetery Hill. Despite heavy losses, the Army of the Potomac held its lines. That evening, Meade reported back to General-in-Chief Henry Halleck, "The enemy attacked me about 4 P.M. this day…and after one of the severest contests of the war was repulsed at all points." Meade ended his message, "I shall remain in my present position to-morrow, but am not prepared to say until better advised of the condition of the army, whether operations will be of an offensive or a defensive character." Having essentially made his decision, Meade summoned his corps commanders and chief of intelligence to assess the condition of the army and to hear from his commanders on courses of action for the next day.
>
> The meeting began around 9 P.M. in which Brigadier General John Gibbon noted, "was at first very informal and in the shape of a conversation." The meeting lasted about two hours as General Meade listened intently to his subordinates' discussion. The tradition in such meetings or council of war is a discussion and then a vote by the officers on the course of action. Meade's Chief of Staff Major General Butterfield posed three questions:
>
> "Under existing circumstances, is it advisable for this army to remain in its present position, or retire to another nearer its base of supplies?
>
> It being determined to remain in present position, shall the army attack or wait the attack of the enemy?
>
> If we wait attack, how long?"
>
> Meade's commanders responded from junior to senior in rank. All wanted to remain on the field another day, but none favored to attack. When the discussion concluded Meade decided that the question was settled and the troops would remain in position. The two-hour discussion and vote formed consensus of the commanders and improved their confidence, resulting in the outcome Meade was seeking—to stay and fight.

Liaison

1-63. Liaison is that contact or intercommunication maintained between elements of military forces and another organization to ensure mutual understanding and unity of purpose and action. Most commonly used for establishing and maintaining close communications, liaison continuously enables direct, physical communications between commands. Commanders use liaison during operations to help facilitate coordination between organizations, de-conflict efforts, and build shared understanding.

1-64. A liaison officer (LNO) represents a commander. A trusted, competent LNO who is properly informed (either a commissioned or a noncommissioned officer) is the key to effective liaison. LNOs must have the commander's full confidence and relevant experience for the mission. The LNO's parent unit or unit of assignment is the sending unit. The unit or activity to which the LNO is sent is the receiving unit. An LNO normally remains at the receiving unit until recalled. LNOs—

- Understand how the commander thinks and interpret the commander's messages.
- Convey the commander's intent, guidance, mission, and concept of operations.
- Represent the commander's position.

(See FM 6-0 for a detailed discussion of the duties and responsibilities of LNOs.)

APPLY CRITICAL AND CREATIVE THINKING

1-65. Thinking includes awareness, perception, reasoning, and intuition. Thinking is naturally influenced by emotion, experience, and bias. As such, commanders and staffs apply critical and creative thinking throughout the operations process to assist them with understanding situations, making decisions, directing actions, and assessing operations.

1-66. Critical thinking is purposeful and reflective thought about what to believe or what to do in response to observations, experiences, verbal or written expressions, or arguments. By thinking critically, individuals formulate judgments about whether the information they encounter is true or false, or if it falls somewhere along a scale of plausibility between true or false. Critical thinking involves questioning information, assumptions, conclusions, and points of view to evaluate evidence, develop understanding, and clarify goals. Critical thinking helps commanders and staffs identify causes of problems, arrive at justifiable conclusions, and make good judgments. Critical thinking helps commanders counter their biases and avoid logic errors.

1-67. Creative thinking examines problems from a fresh perspective to develop innovative solutions. Creative thinking creates new and useful ideas, and reevaluates or combines old ideas to solve problems. Leaders face unfamiliar problems that require new or original approaches to solve them. This requires creativity and a willingness to accept change, newness, and a flexible outlook of new ideas and possibilities.

1-68. Breaking old habits of thought, questioning the status quo, visualizing a better future, and devising responses to new problems require creative thinking. During operations, leaders routinely face unfamiliar problems or old problems under new conditions. Leaders apply creative thinking to gain new insights, novel approaches, fresh perspectives, and new ways of understanding problems and conceiving ways to solve them. (See ATP 5-0.1 for creative thinking tools and techniques.)

1-69. Both critical and creative thinking must intentionally include ethical reasoning—the deliberate evaluation that decisions and actions conform to accepted standards of conduct. Ethical reasoning within critical and creative thinking helps commanders and staffs anticipate ethical hazards and consider options to prevent or mitigate the hazards within their proposed COAs. (See ADP 6-22 for a detailed discussion of ethical reasoning.)

1-70. Commanders may form red teams to help the staff think critically and creatively and to avoid groupthink, mirror imaging, cultural missteps, and tunnel vision. Red teaming enables commanders to explore alternative plans and operations in the context of an OE and from the perspective of unified action partners, adversaries, and others. Throughout the operations process, red team members help clarify the problem and explain how others (unified action partners, the population, and the enemy) potentially view the problem. Red team members challenge assumptions and the analysis used to build the plan. (See JP 5-0 for a detailed discussion of red teams and red teaming.)

INTEGRATING PROCESSES

1-71. Commanders and staffs integrate the warfighting functions and synchronize the force to adapt to changing circumstances throughout the operations process. They use several integrating processes to do this. An integrating process consists of a series of steps that incorporate multiple disciplines to achieve a specific end. For example, during planning, the military decision-making process (MDMP) integrates the commander and staff in a series of steps to produce a plan or order. Key integrating processes that occur throughout the operations process include—

Chapter 1

- Intelligence preparation of the battlefield.
- Information collection.
- Targeting.
- Risk management.
- Knowledge management.

INTELLIGENCE PREPARATION OF THE BATTLEFIELD

1-72. *Intelligence preparation of the battlefield* is the systematic process of analyzing the mission variables of enemy, terrain, weather, and civil considerations in an area of interest to determine their effect on operations (ATP 2-01.3). Led by the intelligence officer, the entire staff participates in IPB to develop and sustain an understanding of the enemy, terrain and weather, and civil considerations. IPB helps identify options available to friendly and threat forces.

1-73. IPB consists of four steps. Each step is performed or assessed and refined to ensure that IPB products remain complete and relevant. The four IPB steps are—
- Define the OE.
- Describe environmental effects on operations.
- Evaluate the threat.
- Determine threat COAs.

IPB begins in planning and continues throughout the operations process. IPB results in intelligence products used to aid in developing friendly COAs and decision points for the commander. Additionally, the conclusions reached and the products created during IPB are critical to planning information collection and targeting. A key aspect of IPB is refinement in preparation and execution. (See ATP 2-01.3 for a detailed discussion of IPB.)

INFORMATION COLLECTION

1-74. *Information collection* is an activity that synchronizes and integrates the planning and employment of sensors and assets as well as the processing, exploitation, and dissemination systems in direct support of current and future operations (FM 3-55). It integrates the functions of the intelligence and operations staffs that focus on answering CCIRs. Information collection includes acquiring information and providing it to processing elements. It has three steps:
- Collection management.
- Task and direct collection.
- Execute collection.

1-75. Information collection helps the commander understand and visualize the operation by identifying gaps in information and aligning reconnaissance, surveillance, security, and intelligence assets to collect information on those gaps. The "decide" and "detect" steps of targeting tie heavily to information collection. (See FM 3-55 for a detailed discussion of information collection to include the relationship between the duties of intelligence and operations staffs.)

TARGETING

1-76. *Targeting* is the process of selecting and prioritizing targets and matching the appropriate response to them, considering operational requirements and capabilities (JP 3-0). Targeting seeks to create specific desired effects through lethal and nonlethal actions. The emphasis of targeting is on identifying enemy resources (targets) that if destroyed or degraded will contribute to the success of the friendly mission. Targeting begins in planning and continues throughout the operations process. The steps of the Army's targeting process are—
- Decide.
- Detect.
- Deliver.
- Assess.

This methodology facilitates engagement of the right target, at the right time, with the most appropriate assets using the commander's targeting guidance.

1-77. Targeting is a multidiscipline effort that requires coordinated interaction among the commander and several staff sections that together form the targeting working group. The chief of staff (executive officer) or the chief of fires (fire support officer) leads the staff through the targeting process. Based on the commander's targeting guidance and priorities, the staff determines which targets to engage and how, where, and when to engage them. The staff then assigns friendly capabilities best suited to produce the desired effect on each target, while ensuring compliance with the rules of engagement. (See ATP 3-60 for a detailed discussion of Army targeting to include how Army targeting nest within the joint targeting cycle.)

RISK MANAGEMENT

1-78. Risk—the exposure of someone or something valued to danger, harm, or loss—is inherent in all operations. Because risk is part of all military operations, it cannot be avoided. Identifying, mitigating, and accepting risk is a function of command and a key consideration during planning and execution. (See chapter 2 for a discussion of risk as an element of operational art.)

1-79. *Risk management* is the process to identify, assess, and control risks and make decisions that balance risk cost with mission benefits (JP 3-0). Commanders and staffs use risk management throughout the operations process to identify and mitigate risks associated with hazards (to include ethical risk and moral hazards) that have the potential to cause friendly and civilian casualties, damage or destroy equipment, or otherwise impact mission effectiveness. Like targeting, risk management begins in planning and continues through preparation and execution. Risk management consists of the following steps:
- Identify hazards.
- Assess hazards.
- Develop controls and make risk decisions.
- Implement controls.
- Supervise and evaluate.

1-80. All staff elements incorporate risk management into their running estimates and provide recommendations to mitigate risk within their areas of expertise. The operations officer coordinates risk management throughout the operations process. (See ATP 5-19 for a detailed discussion of the risk management process.)

KNOWLEDGE MANAGEMENT

1-81. *Knowledge management* is the process of enabling knowledge flow to enhance shared understanding, learning, and decision making (ADP 6-0). It facilitates the transfer of knowledge among commanders, staffs, and forces to build and maintain situational understanding. Knowledge management helps get the right information to the right person at the right time to facilitate decision making. Knowledge management uses a five-step process to create shared understanding. The steps of knowledge management include—
- Assess.
- Design.
- Develop.
- Pilot.
- Implement.

(See ATP 6-01.1 for discussion on knowledge management.)

BATTLE RHYTHM

1-82. Commanders and staffs must integrate and synchronize numerous activities, meetings, and reports within their headquarters, and with higher, subordinate, supporting, and adjacent units as part of the operations process. They do this by establishing the unit's battle rhythm. *Battle rhythm* is a deliberate, daily schedule of command, staff, and unit activities intended to maximize use of time and synchronize staff actions (JP 3-33). A unit's battle rhythm provides structure for managing a headquarters' most important internal resource—the time of the commander and staff. A headquarters' battle rhythm consists of a series of meetings, report requirements, and other activities synchronized by time and purpose. These activities may be daily, weekly, monthly, or quarterly depending on the echelon, type of operation, and planning horizon. An effective battle rhythm—

- Facilitates interaction among the commander, staff, and subordinate commanders.
- Supports building and maintaining shared understanding throughout the headquarters.
- Establishes a routine for staff interaction and coordination.

1-83. There is no standard battle rhythm for every situation. Different echelons, types of units, and types of operations require commanders and staffs to develop a battle rhythm based on the situation. During large-scale ground combat, where lethality and time constraints require rapid planning and decision cycles, the unit's battle rhythm focuses on defeating the enemy. Daily battle rhythm events may consist of a morning and evening current operations update brief, a targeting meeting, and a combined plans and future operations update brief. In operations dominated by stability tasks, where headquarters are often static, the battle rhythm may be more deliberate with daily, weekly, and monthly working groups and boards. While the battle rhythm establishes a routine for a headquarters, the unit's battle rhythm is not fixed. Commanders modify the battle rhythm as the situation evolves. (See ATP 6-0.5 for a detailed discussion of battle rhythm to include examples of common meetings, working groups, and boards.)

Chapter 2
Planning

To be practical, any plan must take account of the enemy's power to frustrate it; the best chance of overcoming such obstruction is to have a plan that can be easily varied to fit the circumstances met; to keep such adaptability, while still keeping the initiative, the best way is to operate along a line which offers alternative objectives.

Sir Basil Henry Liddell Hart

This chapter defines planning and describes the functions of planning and plans. It discusses planning at the levels of warfare, operational art, integrated planning, and key components of a plan. The chapter concludes with guides for effective planning and planning pitfalls to avoid.

FUNDAMENTALS OF PLANNING

2-1. **Planning is the art and science of understanding a situation, envisioning a desired future, and determining effective ways to bring that future about.** Planning helps leaders understand situations; develop solutions to problems; direct, coordinate, and synchronize actions; prioritize efforts; and anticipate events. In its simplest form, planning helps leaders determine how to move from the current state of affairs to a more desirable future state while identifying potential opportunities and threats along the way.

2-2. Planning is a continuous learning activity. While planning may start an iteration of the operations process, planning does not stop with the production of an order. During preparation and execution, the commander and staff continuously refine the order to account for changes in the situation. Subordinates and others provide assessments about what works, what does not work, and how the force can do things better. In some circumstances, commanders may determine that the current order (to include associated branches and sequels) no longer applies. In these instances, instead of modifying the current order, commanders reframe the problem and develop a new plan.

2-3. Planning may be highly structured, involving the commander, staff, subordinate commanders, and others who develop a fully synchronized plan or order. Planning may also be less structured, involving a commander and selected staff who quickly determine a scheme of maneuver for a hasty attack. Sometimes the planned activity is quite specific with very clear goals. At other times, planning must first determine the activity and the goals. Planning is conducted along various planning horizons, depending on the echelon and circumstances. Some units may plan out to years and months, others out to days and hours.

2-4. Planning techniques and methods vary based on circumstances. Planners may plan forward, starting with the present conditions and laying out potential decisions and actions forward in time. Planners also plan in reverse, starting with the envisioned end state and working backward in time to the present. Planning methods may be analytical, as in the MDMP, or more systemic, as in the Army design methodology (ADM).

2-5. A product of planning is a plan or order—a directive for future action. Commanders issue plans and orders to subordinates to communicate their visualization of the operations and to direct action. Plans and orders synchronize the action of forces in time, space, and purpose to achieve objectives and accomplish the mission. They inform others outside the organization on how to cooperate and provide support.

2-6. Plans and orders describe a situation, establish a task organization, lay out a concept of operations, assign tasks to subordinate units, and provide essential coordinating instructions. The plan serves as a foundation for which the force can rapidly adjust from based on changing circumstance. The measure of a good plan is not whether execution transpires as planned, but whether the plan facilitates effective action in the face of unforeseen events.

Chapter 2

2-7. Plans and orders come in many forms and vary in the scope, complexity, and length of time they address. Generally, commanders and staffs develop an operation plan (OPLAN) well in advance of execution; it is not executed until directed. An OPLAN becomes an OPORD when directed for execution based on a specific time or event. A FRAGORD is an abbreviated form of an OPORD issued as needed to change or modify an OPORD during the conduct of operations. (See FM 6-0 for Army formats for plans and orders.)

THE SCIENCE AND ART OF PLANNING

> *Logistics comprises the means and arrangements which work out the plans of strategy and tactics. Strategy decides where to act; logistics brings the troops to this point.*
>
> Antoine Henri de Jomini

2-8. Planning is both a science and an art. Many aspects of military operations, such as movement rates, fuel consumption, and weapons effects, are quantifiable. They are part of the science of planning. The combination of forces, choice of tactics, and arrangement of activities belong to the art of planning. Soldiers often gain knowledge of the science of planning through institutional training and study. They gain understanding of the art of planning primarily through operational training and experience. Effective planners are grounded in both the science and the art of planning.

2-9. The science of planning encompasses aspects of operations that can be measured and analyzed. These aspects include the physical capabilities of friendly and enemy organizations. The science of planning includes a realistic appreciation for time-distance factors; an understanding of how long it takes to initiate certain actions; the techniques and procedures used to accomplish planning tasks; and the terms and graphics that compose the language of military operations. While not easy, the science of planning is fairly straightforward.

2-10. Mastery of the science of planning is necessary for military professionals to understand the physical and procedural constraints under which units operate. These constraints include the effects of terrain, weather, and time on friendly and enemy forces. However—because combat is an intensely human activity—the solution to problems cannot be reduced to a formula. This realization necessitates the study of the art of planning.

2-11. The art of planning requires understanding the dynamic relationships among friendly forces, the threat, and other aspects of an OE during operations. It includes making decisions based on skilled judgment acquired from experience, training, study, imagination, and critical and creative thinking. Commanders apply judgment based on their knowledge and experience to select the right time and place to act, assign tasks, prioritize actions, and allocate resources. The art of planning involves the commander's willingness to accept risk.

2-12. Planning requires creative application of doctrine, units, and resources. It requires a thorough knowledge and application of the fundamentals of unified land operations (see ADP 3-0) and the fundamentals of tactics (see ADP 3-90). The art of planning involves developing plans within the commander's intent and planning guidance by choosing from interrelated options, including—

- Arrangement of activities in time, space, and purpose.
- Assignment of tactical mission tasks and tactical enabling tasks.
- Task organization of available forces and resource allocation.
- Choice and arrangement of control measures.
- Tempo.
- The risk the commander is willing to take.

2-13. These interrelated options define a starting point from which planners create distinct solutions to particular problems. Each solution involves a range of options. Each balances competing demands and requires judgment. The variables of mission, enemy, terrain and weather, troops and support available, time available, and civil considerations (known as METT-TC) always combine to form a different set of circumstances. There are no checklists that adequately apply to every situation.

THE FUNCTIONS OF PLANNING

I tell this story to illustrate the truth of the statement I heard long ago in the Army: Plans are worthless, but planning is everything.

Dwight D. Eisenhower

2-14. Imperfect knowledge and assumptions about the future are inherent in all planning. Planning cannot predict with precision how enemies will react or how civilians will respond during operations. Nonetheless, the understanding and learning that occurs during planning have great value. Even if units do not execute the plan exactly as envisioned—and few ever do—planning results in an improved understanding of the situation that facilitates future decision making. Planning and plans help leaders—

- Understand situations and develop solutions to problems.
- Task-organize the force and prioritize efforts.
- Direct, coordinate, and synchronize action.
- Anticipate events and adapt to changing circumstances.

UNDERSTAND SITUATIONS AND DEVELOP SOLUTIONS TO PROBLEMS

2-15. Planning helps commanders and staffs understand situations to include discerning the relationship of the operational and mission variable. Effective planning not only helps leaders understand the land domain, but it helps leaders understand how capabilities in the air, maritime, space, and cyberspace domains and the information environment impact operations on land and vice versa.

2-16. Understanding the situation requires both analysis and synthesis. Analysis is the process of studying a situation by successively dividing it into parts and addressing each part in turn. For example, the initial stages of mission analysis and IPB rely heavily on analysis. Understanding the parts of a situation is necessary; however, understanding the parts alone does not provide an appreciation of the relationships among the parts. That appreciation requires synthesis. Synthesis is thinking about how the parts of a situation work together as a whole rather than in isolation. As part of planning, the commander and staff synthesize results of mission analysis to make sense of the situation before developing COAs.

2-17. Planning also helps leaders identify problems and develop solutions to solve or manage those problems. Not all problems require the same level of planning. Leaders often identify simple problems immediately and quickly decide on a solution—sometimes on the spot. Planning is critical, however, when a problem is actually a set of interrelated issues, and the solution to each affects the others. For unfamiliar situations, planning offers ways to solve the complete set of problems as a whole. In general, the more complex a situation is, the more important and involved the planning effort becomes.

TASK-ORGANIZE THE FORCE AND PRIORITIZE EFFORTS

2-18. When developing their concept of operations, commanders first visualize the decisive operation that directly accomplishes the mission. They then visualize how shaping and sustaining operations support the decisive operation. The decisive operation prioritizes effort and is the focal point around which the plan is developed. When developing associated tasks to subordinate units, commanders ensure subordinates have the capabilities and resources to accomplish their assigned tasks. They do this by task-organizing the force and establishing priorities of support. Commanders consider the following principles of war when task-organizing the force and prioritizing efforts:

- Mass: concentrate the effects of combat power at the decisive place and time.
- Economy of force: allocate minimum-essential combat power to secondary efforts.
- Unity of command: for every objective, ensure unity of effort under one responsible commander.

2-19. *Task-organizing* is the act of designing a force, support staff, or sustainment package of specific size and composition to meet a unique task or mission (ADP 3-0). It includes providing assets to subordinate commanders and establishing their command and support relationships. Some assets are retained under the commander's immediate control to retain flexibility to exploit opportunities or counter threats.

Chapter 2

2-20. Task-organizing results in a *task organization*—**a temporary grouping of forces designed to accomplish a particular mission.** The unit's task organization is stipulated in the base plan or order or addressed in Annex A (Task Organization) to the base plan or order. The OPLAN or OPORD also stipulates changes in the task organization by phase or event. During execution, commanders modify the task organization as required based on the situation through FRAGORDs. (See FM 6-0 for task organization formats in Army plans and orders.)

2-21. Commanders avoid exceeding the span of control of a subordinate headquarters when task-organizing. Span of control refers to the number of subordinate units under a single commander. This number is situation dependent and may vary. Allocating more units to subordinate commanders gives subordinates greater flexibility and increases options and combinations. However, increasing the number of subordinate units increases the number of decisions the commander must make, and that may decrease agility. Running estimates and COA analysis provide the information that helps commanders determine the best task organization to—

- Facilitate the commander's intent and concept of operations.
- Weight the decisive operation or main effort.
- Create effective combined arms teams.
- Retain flexibility to meet unforeseen events and support future operations.
- Allocate resources with minimum restrictions on their employment.

Army Command and Support Relationships

2-22. Command and support relationships provide the basis for unity of command and are essential to the exercise of mission command. Army command relationships define command responsibility and authority. Army support relationships define the purpose, scope, and effect desired when one capability supports another. Establishing clear command and support relationships is fundamental to organizing for any operation.

2-23. Army command relationships define superior and subordinate relationships between unit commanders. By specifying a chain of command, command relationships unify effort and enable commanders to use subordinate forces with maximum flexibility. Army command relationships include—

- Organic.
- Assigned.
- Attached.
- Operational control.
- Tactical control.

2-24. Army command relationships identify the authorities and degree of control of the gaining Army commander. For example, operational control gives gaining commanders the authority to assign missions and further task-organize forces placed under their operational control. Tactical control on the other hand, gives gaining commanders the authority to assign missions, but not further task-organize forces placed under their tactical control. The type of command relationship often relates to the expected longevity of the relationship between the headquarters involved and quickly identifies the administrative and logistic support that the gaining and losing Army commanders provide.

2-25. A support relationship is established by a superior commander between subordinate commanders when one organization should aid, protect, complement, or sustain another force on a temporary basis. Designating support relationships is an important aspect of mission command in that it provides a flexible means of establishing and changing priorities with minimal additional instruction. Army support relationships are—

- Direct support.
- General support.
- Reinforcing.
- General support-reinforcing.

2-26. Each Army support relationship identifies specific authorities and their responsibilities between the supported and supporting units to include who has the authority to sustain, establish communication with, position, and set priorities for the supporting force. For example, an artillery unit in direct support of a maneuver unit is positioned and has priorities established by the maneuver unit. A sustainment unit in general support of multiple units is positioned and has priorities of support established by its parent unit.

2-27. Establishing clear command and support relationships is fundamental to organizing for any operation. These relationships are doctrinally defined and establish clear responsibilities and authorities between subordinate and supporting units. Knowing the inherent responsibilities of each command and support relationship allows commanders to effectively organize their forces and helps supporting commanders understand their unit's role in the organizational structure. (See ADP 3-0 for a detailed discussion of Army command and support relationships.)

Prioritizing Effort

2-28. In addition to task-organizing, commanders establish priorities of support during planning and shift priorities during execution as the situation requires. A *priority of support* **is a priority set by the commander to ensure a subordinate unit has support in accordance with its relative importance to accomplish the mission.** Priorities of movement, fires, sustainment, and protection all illustrate priorities of support that commanders use to weight the decisive operation or the main effort if the operation is phased. The *main effort* is a designated subordinate unit whose mission at a given point in time is most critical to overall mission success (ADP 3-0). The main effort is weighted with the preponderance of *combat power*—the total means of destructive, constructive, and information capabilities that a military unit or formation can apply at a given time (ADP 3-0). Designating a main effort temporarily gives that unit priority of support. Commanders shift resources and priorities to the main effort as circumstances require. Commanders may shift the main effort several times during an operation. When executed, the unit conducting the decisive operation—the operation that directly accomplishes the mission—is always the main effort.

DIRECT, COORDINATE, AND SYNCHRONIZE ACTIONS

2-29. Plans and orders are the principle means commanders use to direct, coordinate, and synchronize actions. Plans and orders also inform those outside the unit how to cooperate and provide support. Good plans direct subordinates by stating what is required (the task) and why (the purpose); they leave how (the method) up to subordinates. They contain the minimum number of control measures needed to coordinate actions and synchronize the warfighting functions to mass the effects of combat power at the decisive point and time.

2-30. Commanders use control measures to assign responsibilities, coordinate fire and maneuver, and control operations. A *control measure* is a means of regulating forces or warfighting functions (ADP 6-0). Control measures assign responsibilities, coordinate actions between forces, impose restrictions, or establish guidelines to regulate freedom of action. Control measures are essential to coordinating subordinates' actions and are located throughout the plan. Control measures unburden subordinate commanders to conduct operations within their assigned AO without additional coordination.

2-31. Control measures can be permissive (which allows something to happen) or restrictive (which limits how something is done). For example, a *coordinated fire line*—a line beyond which conventional surface-to-surface direct fire and indirect fire support means may fire at any time within the boundaries of the establishing headquarters without additional coordination but does not eliminate the responsibility to coordinate the airspace required to conduct the mission (JP 3-09)—illustrates a permissive control measure. A *route*—the prescribed course to be traveled from a specific point of origin to a specific destination (FM 3-90-1)—illustrates a restrictive control measure. (ADP 1-02 contains definitions and symbols of control measures.)

2-32. Synchronization is the arrangement of military actions in time, space, and purpose. Plans and orders synchronize the warfighting functions to mass the effects of combat power at the chosen place and time. Synchronization is a means of control, not an end. Commanders balance necessary synchronization against desired agility and initiative.

2-33. Overemphasizing the direction, coordination, and synchronization functions of planning may result in detailed and rigid plans that stifle initiative. Mission command encourages the use of mission orders to avoid creating overly restrictive instructions to subordinates. Mission orders direct, coordinate, and synchronize

Chapter 2

actions while allowing subordinates the maximum freedom of action to accomplish missions within the commander's intent. (See paragraphs 2-119 through 2-126 for a discussion on mission orders.)

ANTICIPATE EVENTS AND ADAPT TO CHANGING CIRCUMSTANCES

In general, campaign projects have to be adjusted to conditions (time, weather), the number of the enemy. ...The more one foresees obstacles to his plans, the less one will find of them later in the execution. In a word, everything must be foreseen; find the problems and resolve them.

Frederick the Great

2-34. A fundamental tension exists between the desire to plan far into the future to facilitate preparation and coordination and the fact that the farther into the future the commander plans, the less certain the plan will remain relevant. Given the fundamentally uncertain nature of operations, the object of planning is not to eliminate uncertainty but to develop a framework for action in the midst of such uncertainty. Planning provides an informed forecast of how future events may unfold. It entails identifying and evaluating potential decisions and actions in advance to include thinking through consequences of certain actions. Planning involves thinking about ways to influence the future as well as ways to respond to potential events.

2-35. Planning keeps the force oriented on future objectives despite the requirements of current operations. Anticipatory planning is essential for seizing and retaining the initiative by allowing commanders and staffs to consider potential decisions and actions in advance. Anticipatory planning reduces the time between decisions and actions during execution, especially at higher echelons. While some actions are implemented immediately, others require forethought and preparation. For example, changing the direction of attack may be a relatively simple and immediate matter for a battalion; however, changing the scheme of maneuver for a division, including all its support, is complicated and time consuming. Changing priority of fires at division level may take considerable time if artillery units must reposition. If leaders wait until an event occurs to begin planning and preparing for it, units may not be able to react quickly enough—ceding the initiative to the enemy.

2-36. During execution planners continue to develop or refine options for potential enemy action and friendly opportunities. By anticipating potential events beforehand, planning promotes flexibility and rapid decision making during execution. As a result, the force anticipates events and acts purposefully and effectively before the enemy can act or before situations deteriorate. Several tools are available to the commander and planners to assist in adapting to changing circumstance to include—

- Decision points.
- Branches.
- Sequels.

2-37. A *decision point* is a point in space and time when the commander or staff anticipates making a key decision concerning a specific course of action (JP 5-0). A decision point is associated to actions by the enemy, the friendly force, or the population and tied to a CCIR. Identifying decision points associated to the execution of a branch or sequel is key to effective planning.

2-38. Planners record decision points on a decision support template and associated matrix. A *decision support template* is a combined intelligence and operations graphic based on the results of wargaming that depicts decision points, timelines associated with movement of forces and the flow of the operation, and other key items of information required to execute a specific friendly course of action (JP 2-01.3). The decision support matrix provides text to recap expected events, decision points, and planned friendly actions. It describes where and when a decision must be made if a specific action is to take place. It ties decision points to named areas of interest (known as NAIs), targeted areas of interest (known as TAIs), CCIRs, collection assets, and potential friendly response options. The staff refines the decision support template and matrix as planning progresses and during execution.

2-39. Plans and orders often require adjustment beyond the initial stages of the operations. A *branch* is the contingency options built into the base plan used for changing the mission, orientation, or direction of movement of a force to aid success of the operation based on anticipated events, opportunities, or disruptions caused by enemy actions and reactions (JP 5-0). Branches anticipate situations that require changes to the basic plan. Such situations could result from enemy action, friendly action, or weather. Commanders build flexibility into their plans and orders by developing branches to preserve freedom of action in rapidly changing conditions.

2-40. A *sequel* is the subsequent operation or phase based on the possible outcomes of the current operation or phase (JP 5-0). Sequels are based on outcomes of current operations to include success, stalemate, or defeat. A counteroffensive, for example, is a logical sequel to a defense; an exploitation and pursuit follow successful attacks. Executing a sequel normally begins another phase of an operation, if not a new operation. Commanders consider and develop sequels during planning and revisit them throughout an operation.

PLANNING AND THE LEVELS OF WARFARE

2-41. It is important to understand how Army planning nests with joint planning and how planning differs at the levels of warfare. The *levels of warfare* are a framework for defining and clarifying the relationship among national objectives, the operational approach, and tactical tasks (ADP 1-01). The three levels are strategic, operational, and tactical. There is no hard boundary between levels of warfare, nor fixed echelon responsible for a particular level.

2-42. The levels of warfare focus a headquarters on one of three broad roles—creating strategy; conducting campaigns and major operations; or sequencing battles, engagements, and actions. The levels of warfare correspond to specific levels of responsibility and planning with decisions at one level affecting other levels. They help commanders visualize a logical arrangement and synchronization of operations, allocate resources, and assign tasks to the appropriate command. Among the levels of warfare, planning horizons differ greatly.

STRATEGIC LEVEL

> *War plans cover every aspect of a war, and weave them all into a single operation that must have a single, ultimate objective in which all particular aims are reconciled. No one starts a war—or rather, no one in his sense ought to do so—without first being clear in his mind what he intends to achieve by that war and how he intends to conduct it. The former is its political purpose; the latter its operational objective.*
>
> Carl von Clausewitz

2-43. The *strategic level of warfare* is the level of warfare at which a nation, often as a member of a group of nations, determines national or multinational (alliance or coalition) strategic security objectives and guidance, then develops and uses national resources to achieve those objectives (JP 3-0). The focus at this level is the development of strategy—a foundational idea or set of ideas for employing the instruments of national power in a synchronized and integrated fashion to achieve national and multinational objectives. The strategic level of war is primarily the province of national leadership in coordination with combatant commanders.

2-44. The National Security Council develops and recommends national security policy options for Presidential approval. The President, the Secretary of Defense, and the Chairman of the Joint Chiefs of Staff provide their orders, intent, strategy, direction, and guidance via strategic direction to the military (Services and combatant commands) to pursue national interest. They communicate strategic direction to the military through written documents referred to as strategic guidance. Key strategic guidance documents include—

- *National Security Strategy of the United States.*
- *National Defense Strategy of the United States.*
- *National Military Strategy of the United States.*
- *Joint Strategic Campaign Plan.*
- *Unified Command Plan.*
- *Guidance for Employment of the Force.*
- *Global Force Management Implementation Guidance.*

(See JP 5-0 for a detailed discussion of strategic direction and guidance.)

2-45. Based on strategic guidance, GCCs and staffs—with input from subordinate commands (to include the theater army) and supporting commands and agencies—update their strategic estimates and develop theater

Chapter 2

strategies. A theater strategy is a broad statement of a GCC's long-term vision that bridges national strategic guidance and the joint planning required to achieve national and theater objectives. The theater strategy prioritizes the ends, ways, and means within the limitations established by the budget, global force management processes, and strategic guidance.

Note. Functional combatant commanders also follow this process within their functional areas.

OPERATIONAL LEVEL

2-46. The *operational level of warfare* is the level of warfare at which campaigns and major operations are planned, conducted, and sustained to achieve strategic objectives within theaters or other operational areas (JP 3-0). Operational-level planning focuses on developing plans for campaigns and other joint operations. A *campaign plan* is a joint operation plan for a series of related major operations aimed at achieving strategic or operational objectives within a given time and space (JP 5-0). Joint force commanders (combatant commanders and their subordinate joint task force commanders) and their component commanders (Service and functional) conduct operational-level planning. Planning at the operational level requires operational art to integrate ends, ways, and means while balancing risk. Operational-level planners use operational design and the joint planning process to develop campaign plans, OPLANs, OPORDs, and supporting plans. (JP 5-0 discusses joint planning. JP 3-31 discusses operational-level planning from a land component perspective.)

2-47. The combatant command campaign plan (CCP) operationalizes the GCC's strategy by organizing and aligning operations and activities with resources to achieve objectives in an area of responsibility. The CCP provides a framework within which the GCC conducts security cooperation activities and military engagement with regional partners. The CCP contains contingency plans that are viewed as branches within the campaign. Contingency plans identify how the command might respond in the event of a crisis. Contingency plans are often phased and have specified end states that seek to re-establish conditions favorable to the United States. Contingency plans have an identified military objective and termination criteria. They may address limited contingency operations or large-scale combat operations.

2-48. The theater army develops a support plan to the CCP. This support plan includes methods to achieve security cooperation, training and exercise programs, and ongoing Army activities within the theater including intelligence, air and missile defense, sustainment, and communications. The theater army also develops supporting plans for contingencies identified by the GCC. These include OPLANs for large-scale ground combat, noncombatant evacuation operations, humanitarian assistance and disaster relief, and other crises response activities. Theater army planners routinely develop, review, and update supporting plans to numbered OPLANs to ensure they remain feasible. This includes a review of Army force structure as well as time-phased force and deployment data.

2-49. Corps and below Army units normally conduct Army tactical planning. However, corps and divisions serving as the base headquarters for a joint task force or land component headquarters employ joint planning and develop joint formatted plans and orders. Corps or divisions receive joint formatted plans and orders when directly subordinate to a joint task force or joint land component command. It is important for these headquarters to be familiar with joint Adaptive Planning and Execution. (See CJCSM 3130.03A for joint formats for plans and orders.) Figure 2-1 illustrates the links among the levels of warfare using military actions in the Gulf War of 1991.

Planning

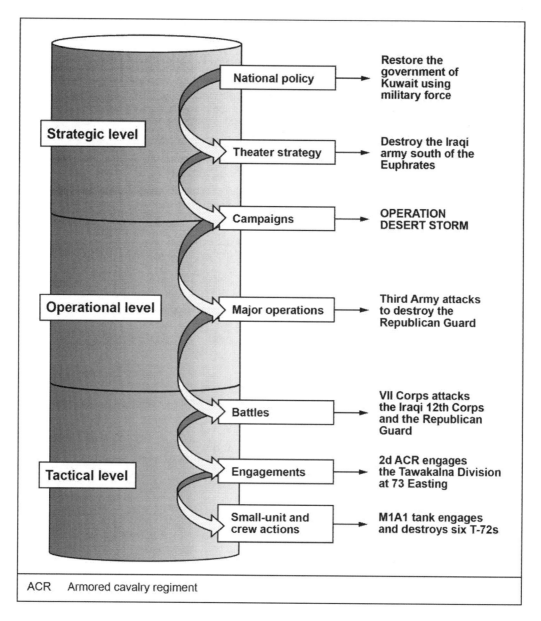

Figure 2-1. Levels of warfare

TACTICAL LEVEL

2-50. The *tactical level of warfare* is the level of warfare at which battles and engagements are planned and executed to achieve military objectives assigned to tactical units or task forces (JP 3-0). Tactical-level planning revolves around how best to achieve objectives and accomplish tasks assigned by higher headquarters. Planning horizons for tactical-level planning are relatively shorter than planning horizons for operational-level planning. Tactical-level planning works within the framework of an operational-level plan and is addressed in Service doctrine or, in the case of multinational operations, the lead nation's doctrine. Army tactical planning is guided by the MDMP for units with a staff and TLP for small-units without a staff.

2-51. Operational- and tactical-level planning complement each other but have different aims. Operational-level planning involves broader dimensions of time, space, and purpose than tactical-level planning involves. Operational-level planners need to define an operational area, estimate required forces, and evaluate requirements. In contrast, tactical-level planning proceeds from an existing operational design. Normally, AOs are prescribed, objectives and available forces are identified, and a general sequence of activities is specified for tactical-level commanders.

OPERATIONAL ART

Nothing succeeds in war except in consequence of a well prepared plan.

Napoleon Bonaparte

2-52. *Operational art* is the cognitive approach by commanders and staffs—supported by their skill, knowledge, experience, creativity, and judgment—to develop strategies, campaigns, and operations to organize and employ military forces by integrating ends, ways, and means (JP 3-0). Operational art applies to all types and aspects of operations. It integrates ends, ways, and means while accounting for risk. Applying operational art requires commanders to answer the following questions:

- What conditions, when established, constitute the desired end state (ends)?
- How will the force achieve these desired conditions (ways)?
- What sequence of actions helps attain these conditions (ways)?
- What resources are required to accomplish that sequence of actions (means)?
- What risks are associated with that sequence of actions and how can they be mitigated (risks)?

Operational art encompasses all levels of warfare. It requires creative vision, broad experience, and a knowledge of capabilities, tactics, and techniques across multiple domains. Commanders and staffs employ operational art during ADM and the MDMP.

OPERATIONAL APPROACH

2-53. It is through operational art that commanders develop and translate their operational approach—a description of the broad actions required to achieve the end state—into a concept of operations. An operational approach is the result of the commander's visualization of what needs to be done in broad terms to solve identified problems. It is the main idea that informs detailed planning. When developing an operational approach, commanders consider ways to employ a combination of defeat mechanisms and stability mechanisms. Defeat mechanisms relate to offensive and defensive operations; stability mechanisms relate to stability operations.

2-54. Army forces use combinations of four defeat mechanisms: destroy, dislocate, disintegrate, and isolate. Destroy means to physically render an enemy force combat-ineffective until it is reconstituted. Dislocate means to employ forces to obtain a significant positional advantage, rendering the enemy's disposition less valuable or irrelevant. Disintegrate means to disrupt the enemy's command and control system, degrading their ability to conduct operations while leading to the enemy's rapid collapse or will to fight. Isolate means to seal off—both physically and psychologically—an enemy from sources of support.

2-55. Applying more than one defeat mechanism simultaneously produces complementary and reinforcing effects not attainable with a single mechanism. Used individually, a defeat mechanism achieves results relative to the amount of effort expended. Using defeat mechanisms in combination creates enemy dilemmas that magnify their effects significantly. Operational art formulates the most effective, efficient way to apply defeat mechanisms. Physically defeating the enemy deprives enemy forces of the ability to achieve those aims. Temporally defeating the enemy anticipates enemy reactions and nullifies them before they can become effective. Cognitively defeating the enemy disrupts decision making and erodes the enemy's will to fight.

2-56. As with defeat mechanisms, combinations of stability mechanisms produce complementary and reinforcing effects that accomplish the mission more effectively and efficiently than single mechanisms do alone. The four stability mechanisms are compel, control, influence, and support. Compel means to use, or threaten to use, lethal force to establish control and dominance, affect behavioral change, or enforce compliance with mandates, agreements, or civil authority. Control involves imposing civil order. Influence means to alter the opinions, attitudes, and ultimately the behavior of foreign friendly, neutral, adversary, and enemy audiences through messages, presence, and actions. Support establishes, reinforces, or sets conditions necessary for the instruments of national power to function effectively.

Planning

ELEMENTS OF OPERATIONAL ART

2-57. In applying operational art, commanders and their staffs use a set of intellectual tools known as the elements of operational art. These tools help commanders understand, visualize, and describe operations and help to formulate their commander's intent and planning guidance to include the operational approach. Commanders may use these tools in any operation; however, their application is broadest in the context of long-term operations.

2-58. Not all elements of operational art apply at all levels of warfare. A battalion commander may be concerned about the tempo of an upcoming operation but is probably not concerned with an enemy's center of gravity. A corps commander may consider all elements of operational art in developing a plan. The application of specific elements of operational art depends on the situation and echelon.

Elements of operational art
- End state and conditions
- Centers of gravity
- Decisive points
- Lines of operations and lines of effort
- Tempo
- Phasing and transitions
- Operational reach
- Culmination
- Basing
- Risk

End State and Conditions

2-59. A central aspect of planning is determining the operation's end state. The *end state* is the set of required conditions that defines achievement of the commander's objectives (JP 3-0). A condition is a reflection of the existing state of an OE. Thus, a desired condition is a sought-after change to an OE. Since every operation should focus on a clearly defined and attainable end state, accurately describing conditions that represent success is important.

2-60. Commanders explicitly describe end state conditions in their planning guidance to shape the development of an operational approach and COAs. Commanders summarize the operation's end state in their commander's intent. A clearly defined end state promotes unity of effort, facilitates integration and synchronization of the force, and guides subordinate initiative during execution.

2-61. Commanders ensure their end state is nested with their higher headquarters' end state and the overall end state for the joint operation. Subordinate operations within the larger plan often have an end state for that particular operation. In these instances commanders often address conditions for transition beyond the current operation to facilitate follow-on operations or an exploitation.

Centers of Gravity

> *For Alexander, Gustavus Adolphus, Charles XII, and Frederick the Great, the center of gravity was their army. If the army had been destroyed, they would all have gone down in history as failures.*
>
> Carl von Clausewitz

2-62. A *center of gravity* is the source of power that provides moral or physical strength, freedom of action, or will to act (JP 5-0). The loss of a center of gravity can ultimately result in defeat. Centers of gravity are not limited to military forces and can be either physical or moral. Physical centers of gravity, such as a capital city or military force, are tangible and typically easier to identify, assess, and target than moral centers of gravity. Forces can often influence physical centers of gravity solely by military means. In contrast, moral centers of gravity are intangible and more difficult to influence; they exist in the cognitive dimension of an information environment. They can include a charismatic leader, powerful ruling elite, or the will of a population.

2-63. As an element of operational art, a center of gravity analysis helps commanders and staffs understand friendly and enemy sources of strength and weakness. This understanding helps to determine ways to undermine enemy strengths by exploiting enemy vulnerabilities while protecting friendly vulnerabilities from enemies attempting to do the same. Understanding friendly and enemy centers of gravity helps the commander and staffs identify decisive points and determine an operational approach to achieve the end state. (See JP 5-0 for more detailed discussions of center of gravity analysis.)

Chapter 2

Decisive Points

2-64. A *decisive point* is a geographic place, specific key event, critical factor, or function that, when acted upon, allows commanders to gain a marked advantage over an enemy or contribute materially to achieving success (JP 5-0). Identifying decisive points helps commanders to select clear, conclusive, attainable objectives that directly contribute to achieving the end state. Geographic decisive points can include port facilities, distribution networks and nodes, and bases of operation. Specific events and elements of an enemy force may also be decisive points. Examples of such events include commitment of an enemy operational reserve and reopening a major oil refinery.

2-65. A common characteristic of decisive points is their importance to a center of gravity. Decisive points are not centers of gravity; they are key to attacking or protecting centers of gravity. A decisive point's importance may cause the enemy to commit significant resources to defend it. The loss of a decisive point weakens a center of gravity and may expose more decisive points, eventually leading to an attack on the center of gravity itself.

2-66. Generally, more decisive points exist in a given operational area than available forces and capabilities can attack, seize, retain, control, or protect. Accordingly, planners study and analyze decisive points and determine which offer the best opportunity to attack the enemy's center of gravity, extend friendly operational reach, or enable the application of friendly forces and capabilities. The art of planning includes selecting decisive points that best lead to establishing end state conditions in a sequence that most quickly and efficiently leads to mission success.

2-67. Decisive points identified for action become objectives. An objective can be physical (an enemy force or a terrain feature) or conceptual as a goal (established rule of law). In the physical sense, an objective is a location on the ground used to orient operations, phase operations, facilitate changes of direction, and provide for unity of effort. In the conceptual sense, an *objective* is the clearly defined, decisive, and attainable goal toward which an operation is directed (JP 5-0). Objectives provide the basis for determining tasks to subordinate units. The most important objective forms the basis for developing the decisive operation. Combined with end state conditions, objectives form the building blocks for developing lines of operations and lines of effort.

Lines of Operations and Lines of Effort

> *If the art of war consists in bringing into action upon the decisive point of the theater of operations the greatest possible force, the choice of the line of operations, being the primary means of attaining this end, may be regarded as the fundamental idea in a good plan of a campaign.*
>
> Antoine Henri de Jomini

2-68. Lines of operations and lines of effort link objectives in time, space, and purpose to achieve end state conditions as shown in figure 2-2. A line of operations links a base of operations to physical objectives which links to end state conditions. Lines of effort link tasks with goal-oriented objectives that focus toward establishing end state conditions. Commanders describe an operation along lines of operations, lines of effort, or a combination of both in their operational approach. Commanders may designate one line as decisive and others as shaping.

Planning

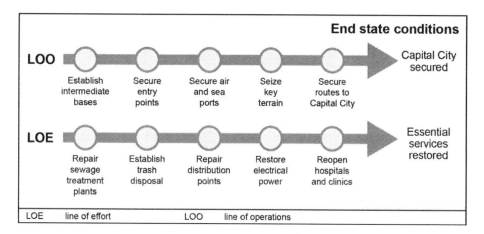

Figure 2-2. Sample line of operations and line of effort

2-69. A *line of operations* is a line that defines the directional orientation of a force in time and space in relation to the enemy and links the force with its base of operations and objectives (ADP 3-0). Lines of operations connect a series of intermediate objectives that lead to control of a geographic or force-oriented objective. Operations designed using lines of operations generally consist of a series of actions executed according to a well-defined sequence.

2-70. Lines of operations can be categorized as interior and exterior. The choice of using interior or exterior lines supports a concept based on the length of movement and the supporting lines of sustainment. Staffs choose interior lines based on the fact that lines of movement and sustainment within an enclosed area are shorter than those lines outside the enclosed area. Interior lines are lines on which a force operates when its operations diverge from a central point. Interior lines allow commanders to move quickly against enemy forces along shorter lines of operations.

2-71. Exterior lines are lines on which a force operates when its operations converge on the enemy. This requires the attacking force to be stronger or more mobile than the enemy. Exterior lines allow commanders to concentrate forces against multiple positions on the ground, thus presenting multiple dilemmas to the enemy. Exterior lines facilitate seizing opportunities to encircle and destroy the weaker or less mobile enemy. While commanders operating on interior lines have the opportunity to set the width of the battlefield, commanders operating on exterior lines have set the disposition of their force to deploy them outside their boundaries.

2-72. A *line of effort* is a line that links multiple tasks using the logic of purpose rather than geographical reference to focus efforts toward establishing a desired end state (ADP 3-0). Lines of effort are essential to long-term planning when positional references to an enemy or adversary have little relevance. In operations involving many nonmilitary factors, lines of effort may be the only way to link tasks to the end state. Lines of effort often enable commanders to visualize how military capabilities can support the other instruments of national power.

Tempo

2-73. Commanders and staff consider tempo both when planning and executing operations. *Tempo* is the relative speed and rhythm of military operations over time with respect to the enemy (ADP 3-0). It reflects the rate of military action. Controlling tempo helps commanders keep the initiative during combat operations or rapidly establish a sense of normalcy during humanitarian crises. During large-scale ground combat, commanders seek to maintain a higher tempo than the enemy does; a rapid tempo can overwhelm an enemy's ability to counter friendly actions. During other operations dominated by stability operations tasks, commanders act quickly to control events and deny the enemy positions of advantage. By acting faster than the situation deteriorates, commanders can change the dynamics of a crisis and restore stability.

Chapter 2

2-74. Several factors affect tempo including the friendly force's status, composition, and mobility. Terrain and weather are other factors. Planning also can accelerate tempo by anticipating decisions and actions in advance. This emphasis on increased tempo, while a guiding principle, is not an unbending rule. Commanders weigh the advantages of acting more quickly against the advantages of preparing more thoroughly.

Phasing and Transitions

2-75. Planning determines the sequence of actions—including the phases and transitions—that best accomplishes the mission. Ideally, commanders plan to accomplish a mission with simultaneous actions throughout the AO. However, operational reach, resource constraints, and the size of the friendly force limits what units can do at one time. In these cases, commanders phase operations. Phasing provides a way to view and conduct operations in manageable parts.

2-76. A *phase* is a planning and execution tool used to divide an operation in duration or activity (ADP 3-0). Within a phase, a large portion of the force executes similar or mutually supporting activities. Achieving a specified condition or set of conditions typically marks the end of a phase. No standard phasing model exists for Army operations. Commanders phase operations as required by the specific circumstances of the problem they are trying to solve.

2-77. A change in phase usually involves a change of mission, task organization, or rules of engagement. Phasing helps in planning and controlling operations during execution. Phasing may be indicated by time, distance, terrain, or an event. Well-designed phases—
- Focus effort.
- Concentrate combat power in time and space at a decisive point.

2-78. Transitions mark a change of focus between phases or between the ongoing operation and execution of a branch or sequel. Shifting priorities among the offense, defense, and stability also involves transitions. Transitions require planning and preparation so the force can maintain the initiative and tempo of operations. Forces are vulnerable during transitions, so commanders establish clear conditions for their execution. Planning identifies potential transitions and accounts for them throughout execution. Effective commanders consider the time required to plan for and execute transitions. Assessment helps commanders measure progress toward such transitions and take appropriate actions to execute them.

Operational Reach

2-79. While planning operations, it is critical to consider *operational reach*—the distance and duration across which a force can successfully employ military capabilities (JP 3-0). The concept of operational reach is inextricably tied to the concept of basing and lines of operations. Although geography may constrain or limit reach, units may extend reach by forward positioning capabilities and resources (such as long-range fires) and leveraging host-nation support. Commanders and staffs consider ways to increase their operational reach in each warfighting function to include leveraging joint and multinational capabilities across all domains. For example, requesting and integrating joint intelligence, reconnaissance, and surveillance in combination with joint fires can significantly increase the unit's operational reach. This requires commanders and staffs to understand and interface with joint planning and processes such as the joint targeting cycle. Commanders and staffs also consider phasing operations based on operational reach.

Culmination

2-80. The limit of a unit's operational reach is its culminating point. The *culminating point* is the point at which a force no longer has the capability to continue its form of operations, offense or defense (JP 5-0). Culmination represents a crucial shift in relative combat power. It is relevant to both attackers and defenders at each level of warfare. While conducting offensive tasks, the culminating point occurs when the force cannot continue the attack and must assume a defensive posture or execute an operational pause. While conducting defensive tasks, it occurs when the force can no longer defend itself and must withdraw or risk destruction. The culminating point is more difficult to identify when Army forces conduct stability tasks. Two conditions can result in culmination: units being too dispersed to achieve security and units lacking required resources to achieve the end state.

Basing

2-81. Basing is an indispensable part of operational art and linked to lines of operations and operational reach. Determining the location and sequence of establishing bases and base camps is essential for projecting power and sustaining the force. Basing may be joint or single Service and will routinely support both U.S. and multinational forces as well as interagency partners. Commanders designate a specific area as a base or base camp and assign responsibility to a single commander for protection, terrain management, and day-to-day operations. (See JP 3-34 and ATP 3-37.10 for more information on basing, bases, and base camps.)

Risk

It is my experience that bold decisions give the best promise of success. But one must differentiate between strategical or tactical boldness and a military gamble. A bold operation is one in which success is not a certainty but which in case of failure leaves one with sufficient forces in hand to cope with whatever situation may arise. A gamble, on the other hand, is an operation which can lead either to victory or to the complete destruction of one's force.

Field Marshal Erwin Rommel

2-82. Risk, uncertainty, and chance are inherent in all military operations. Success during operations depends on a willingness to identify, mitigate, and accept risk to create opportunities. When considering how much risk to accept with a COA, commanders consider risk to the force and risk to the mission. Commanders need to balance the tension between protecting the force, and accepting risks that must be taken to accomplish their mission. They apply judgment with regard to the importance of an objective, time available, and anticipated cost.

2-83. Mission command requires that commanders and subordinates accept risk, exercise initiative, and act decisively, even when the outcome is uncertain. Commanders focus on creating opportunities rather than simply preventing defeat—even when preventing defeat appears safer. Reasonably estimating and intentionally accepting risk is not gambling. Gambling is making a decision in which the commander risks the force without a reasonable level of information about the outcome. Therefore, commanders avoid gambles. Commanders carefully determine risks, analyze and minimize as many hazards as possible, and then accept risk to accomplish the mission.

2-84. Inadequate planning and preparation puts forces at risk, as does delaying action while waiting for perfect intelligence and synchronization. Reasonably estimating and intentionally accepting risk is fundamental to successful operations. Experienced commanders balance audacity and imagination against risk and uncertainty to strike in a manner, place, and time unexpected by enemy forces. This is the essence of surprise. Planning should identify risks to mission accomplishment. Part of developing an operational approach includes answering the question, "What is the chance of failure or unacceptable consequences in employing the operational approach?" Risks range from resource shortfalls to an approach that alienates a population. Identified risks are communicated to higher headquarters, and risk mitigation guidance is provided in the commander's planning guidance.

INTEGRATED PLANNING

2-85. Planning activities occupy a continuum ranging from conceptual to detailed as shown in figure 2-3 on page 2-16. Understanding an OE and its problems, determining the operation's end state, establishing objectives, and sequencing the operation in broad terms all illustrate conceptual planning. Conceptual planning generally corresponds to the art of operations and is the focus of a commander with staff support. The commander's activities of understanding and visualizing are key aspects of conceptual planning.

Chapter 2

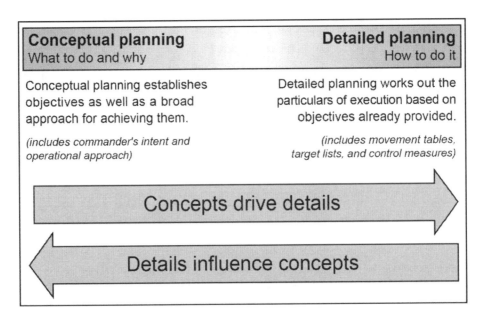

Figure 2-3. Integrated planning

2-86. Detailed planning translates the broad operational approach into a complete and practical plan. Generally, detailed planning is associated with aspects of science, such as movement tables, fuel consumption, target list, weapon effects, and time-distance factors. Detailed planning falls under the purview of the staff, focusing on specifics of execution. Detailed planning works out the scheduling, coordination, or technical problems involved with moving, sustaining, synchronizing, and directing the force. Detailed planning does not mean developing plans with excessive control measures that impede subordinate freedom of action. Planners develop mission orders that establish those controls necessary to coordinate and synchronize the force as a whole. They leave much of the *how* to accomplish tasks to the subordinate.

2-87. The commander personally leads the conceptual component of planning. While they are engaged in parts of detailed planning, commanders leave most specifics to the staff. Conceptual planning provides the basis for all subsequent planning. The commander's intent and operational approach provide the framework for the entire plan. This framework leads to a concept of operations and associated schemes of support, such as schemes of intelligence, maneuver, fires, protection, and sustainment. In turn, the schemes of support lead to the specifics of execution, including tasks to subordinate units and detailed annexes to the OPLAN or OPORD. However, the dynamic does not operate in only one direction. Conceptual planning must respond to detailed constraints. For example, the realities of a deployment schedule (a detailed concern) influence the operational approach (a conceptual concern).

2-88. Successful planning requires the integration of both conceptual and detailed thinking. Army leaders employ several methodologies for planning, determining the appropriate mix based on the scope of the problem, time available, and availability of a staff. Planning methodologies include—
- Army design methodology.
- The military decision-making process.
- Troop leading procedures.
- Rapid decision-making and synchronization process.
- Army problem solving.

ARMY DESIGN METHODOLOGY

2-89. **Army design methodology is a methodology for applying critical and creative thinking to understand, visualize, and describe problems and approaches to solving them.** ADM is particularly useful as an aid to conceptual planning, but it must be integrated with the detailed planning typically associated with the MDMP to produce executable plans and orders. There is no one way or prescribed set of steps to employ the ADM. There are, however, several activities associated with ADM including framing an OE, framing problems, developing an operational approach, and reframing when necessary as shown in figure 2-4. While planners complete some activities before others, the understanding and learning within one activity may require revisiting the learning from another activity. Thus, ADM is iterative in nature.

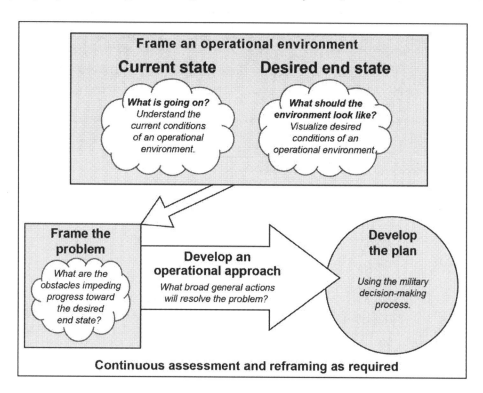

Figure 2-4. Activities of Army design methodology

2-90. When problems are difficult to identify, the operation's end state is unclear, or a COA is not self-evident, commanders employ ADM. This is often the case when developing long-range plans for extended operation or developing supporting plans to the CCP and associated contingencies. The results of ADM include an understanding of an OE and problem, the initial commander's intent, and an operational approach that serves as the link between conceptual and detailed planning. Based on their understanding and learning gained during ADM, commanders issue planning guidance—to include an operational approach—to guide more detailed planning using the MDMP. (See ATP 5-0.1 for techniques for employing ADM.)

2-91. ADM includes interconnected thinking activities that aid in conceptual planning. By first framing an OE and its associated problems, ADM helps commanders and staffs to think about the situation in depth. This in-depth thinking enables them to develop a more informed approach to solve or manage identified problems. During execution, ADM supports organizational learning through reframing. A reframe is a shift in understanding that leads to a new perspective on the problem or its resolution. Reframing is the activity of revisiting earlier hypotheses, conclusions, and decisions that underpin the current operational approach. In essence, reframing reviews what the commander and staff believe they understand about an OE, the problem, and the desired end state.

THE MILITARY DECISION-MAKING PROCESS

2-92. The ***military decision-making process*** **is an iterative planning methodology to understand the situation and mission, develop a course of action, and produce an operation plan or order.** It is an orderly, analytical process that integrates the activities of the commander, staff, and subordinate headquarters in the development of a plan or order. The MDMP helps leaders apply thoroughness, clarity, sound

Chapter 2

judgement, logic, and professional knowledge to develop situational understanding and produce a plan or order that best accomplishes the mission.

2-93. The MDMP consists of seven steps. Each step of the MDMP has inputs, a series of sub-steps, and outputs. The outputs lead to an increased understanding of the situation facilitating the next step of the MDMP. Commanders and staffs generally perform these steps sequentially; however, before producing the plan or order, they may revisit several steps in an iterative fashion as they learn more about the situation. The seven steps are—

- Step 1 – Receipt of mission.
- Step 2 – Mission analysis.
- Step 3 – COA development.
- Step 4 – COA analysis.
- Step 5 – COA comparison.
- Step 6 – COA approval.
- Step 7 – Orders production, dissemination, and transition.

2-94. Commanders initiate the MDMP upon receipt of, or in anticipation of, a mission. Commanders and staffs often begin planning in the absence of an approved higher headquarters' OPLAN or OPORD. In these instances, they start planning based on a warning order (WARNORD), a planning order, or an alert order from higher headquarters. This requires active collaboration with the higher headquarters and parallel planning among echelons as the plan or order is developed. (See FM 6-0 for detailed instructions for conducting the MDMP.)

TROOP LEADING PROCEDURES

2-95. The MDMP and TLP are similar but not identical. ***Troop leading procedures are a dynamic process used by small-unit leaders to analyze a mission, develop a plan, and prepare for an operation.*** TLP extend the MDMP to the small-unit level. Commanders with a coordinating staff use the MDMP as their primary planning process. Company-level and smaller units lack formal staffs and use TLP to plan and prepare for operations. This places the responsibility for planning primarily on the commander or small-unit leader with assistance from forward observers, supply sergeants, and other specialists in the unit.

2-96. TLP enable small-unit leaders to maximize available planning time while developing effective plans and preparing their units for an operation. TLP consist of eight steps. The sequence of the steps of TLP is not rigid. Leaders modify the sequence to meet the mission, situation, and available time. Leaders may perform some steps concurrently while performing other steps continuously throughout the operation. The eight steps are—

- Step 1 – Receive the mission.
- Step 2 – Issue a warning order.
- Step 3 – Make a tentative plan.
- Step 4 – Initiate movement.
- Step 5 – Conduct reconnaissance.
- Step 6 – Complete the plan.
- Step 7 – Issue the order.
- Step 8 – Supervise and refine.

2-97. Leaders use TLP when working alone or with a small group to solve tactical problems. For example, a company commander may use the executive officer, first sergeant, fire support officer, supply sergeant, and communications sergeant to assist during TLP. (See FM 6-0 for a detailed discussion on conducting TLP.)

RAPID DECISION-MAKING AND SYNCHRONIZATION PROCESS

2-98. The rapid decision-making and synchronization process (RDSP) is a decision-making and planning technique that commanders and staffs commonly use during execution when available planning time is limited. While the MDMP seeks an optimal solution, the RDSP seeks a timely and effective solution within the commander's intent. Using the RDSP lets leaders avoid the time-consuming requirements of developing

Planning

decision criteria and multiple COAs. Under the RDSP, leaders combine their experiences and intuition to quickly understand the situation and develop a COA. The RDSP is based on an existing order and the commander's priorities as expressed in the order. The RDSP includes five steps:
- Step 1 – Compare the current situation to the order.
- Step 2 – Determine that a decision, and what type, is required.
- Step 3 – Develop a course of action.
- Step 4 – Refine and validate the course of action.
- Step 5 – Issue the implement the order.

(See chapter 4 for a more detailed discussion decision making during execution and the RDSP.)

ARMY PROBLEM SOLVING

2-99. The ability to recognize and effectively solve problems is an essential skill for Army leaders. Where the previous methodologies are designed for planning operations, Army problem solving is a methodology available for leaders in identifying and solving a variety of problems. Similar in logic to the MDMP, Army problem solving is an analytical approach to defining a problem, developing possible solutions to solve the problem, arriving at the best solution, developing a plan, and implementing that plan to solve the problem. The steps to Army problem solving are—
- Step 1 – Gather information.
- Step 2 – Identify the problem.
- Step 3 – Develop criteria.
- Step 4 – Generate possible solutions.
- Step 4 – Analyze possible solutions.
- Step 6 – Compare possible solutions.
- Step 7 – Make and implement the decision.

(See FM 6-0 for a detailed discussion of Army problem solving.)

KEY COMPONENTS OF A PLAN

An order should not trespass on the province of a subordinate. It should contain everything which is beyond the independent authority of the subordinate, but nothing more.

Field Service Regulations (1905)

2-100. The mission statement, commander's intent, and concept of operations are key components of a plan that serve as the framework for an operation. Commanders ensure their mission and commander's intent nest with those of their higher headquarters. While the commander's intent focuses on the end state, the concept of operations focuses on the way or sequence of actions by which the force will achieve the end state. The concept of operations expands on the mission statement and commander's intent. Within the concept of operations, commanders establish objectives as intermediate goals toward achieving the operation's end state.

MISSION STATEMENT

2-101. The *mission* is the task, together with the purpose, that clearly indicates the action to be taken and the reason therefore (JP 3-0). Commanders analyze a mission based on their higher commander's intent, specified tasks, and implied tasks. Results of that analysis yield the essential task—the task that when executed accomplishes the mission. The essential task becomes the "what" of the mission statement—a clear statement of the action to be taken and the reason for taking it. The five elements of a mission statement answer these questions:
- Who will execute the operation (unit or organization)?
- What is the unit's essential task (normally a tactical mission task or tactical enabling task)?
- Where will the operation occur (AO, objective, engagement areas, or grid coordinates)?
- When will the operation begin (by time or event)?
- Why will the force conduct the operation (for what purpose)?

Chapter 2

2-102. The "who," "where," and "when" of a mission statement are straightforward. The "what" and "why" are more challenging to write and can confuse subordinates if not stated clearly. The "what" is a task and is expressed in terms of action verbs. (See ADP 3-90 for a list of tactical mission tasks.) These tasks are defined and measurable and can be grouped as "actions by friendly forces" or "effects on enemy forces." The "why" puts the task into context by describing the reason for performing it. The mission's purpose facilitates initiative in changing circumstances.

COMMANDER'S INTENT

2-103. The commander's intent succinctly describes what constitutes success for the operation. It includes the operation's purpose, key tasks, and conditions that define the end state. When describing the purpose of the operation, the commander's intent does not restate the "why" of the mission statement. Rather, it describes the broader purpose of the unit's operation in relationship to the higher commander's intent and concept of operations.

2-104. *Key tasks* are those activities the force must perform as a whole to achieve the desired end state (ADP 6-0). During execution—when significant opportunities present themselves or the concept of operations no longer fits the situation—subordinates use key tasks to keep their efforts focused on achieving the desired end state. Examples of key tasks include terrain the force must control or an effect the force must have on the enemy.

2-105. The end state is a set of desired future conditions the commander wants to exist when an operation ends. Commanders describe the operation's end state by stating the desired conditions of the friendly force in relationship to desired conditions of the enemy, terrain, and civil considerations. A clearly defined end state promotes unity of effort among the force and with unified action partners.

CONCEPT OF OPERATIONS

2-106. **The *concept of operations* is a statement that directs the manner in which subordinate units cooperate to accomplish the mission and establishes the sequence of actions the force will use to achieve the end state.** The concept of operations describes how the commander sees the actions of subordinate units fitting together to accomplish the mission. At a minimum, it includes a scheme of maneuver and scheme of fires. Where the commander's intent focuses on the end state, the concept of operations focuses on the method by which the operation uses and synchronizes the warfighting functions to translate the vision and end state into action.

2-107. The concept of operations describes the combination of offensive, defensive, or stability operations and how these tasks complement each other. It describes the deep, close, support, and consolidation areas; decisive, shaping, and sustaining operations within those areas; and main and supporting efforts.

2-108. In developing the concept of operations, commanders and staffs ensure their concepts nest with that of their higher headquarters. ***Nested concepts* is a planning technique to achieve unity of purpose whereby each succeeding echelon's concept of operations is aligned by purpose with the higher echelons' concept of operations.** An effective concept of operations describes how the forces will support a mission of the higher headquarters and how the actions of subordinate units fit together to accomplish a mission.

2-109. The operations overlay—part of Annex C (Operations) to an OPLAN or OPORD—supplements the concept of operations by depicting graphic control measures used to direct operations. A *graphic control measure* is a symbol used on maps and displays to regulate forces and warfighting functions (ADP 6-0). Graphic control measures include symbols for boundaries, fire support coordination measures, some airspace control measures, air defense areas, and obstacles. Commanders establish them to regulate maneuver, movement, airspace use, fires, and other aspects of operations. (See ADP 1-02 for instructions depicting graphic control measures.)

GUIDES TO EFFECTIVE PLANNING

Now the general who wins a battle makes many calculations in his temple ere the battle is fought.

Sun Tzu

2-110. Planning is an inherent and fundamental part of command and control, and commanders are the single most important factor in effective planning. Effective planning requires dedication, study, and practice. Planners must be technically and tactically competent within their areas of expertise and understand basic planning concepts. The following aids in effective planning:
- Incorporate the tenets of unified land operations.
- Commanders focus planning.
- Develop simple, flexible plans through mission orders.
- Optimize available planning time.
- Focus on the right planning horizon.
- Determine relevant facts and develop assumptions.

INCORPORATE THE TENETS OF UNIFIED LAND OPERATIONS

2-111. *Tenets of operations* are desirable attributes that should be built into all plans and operations and are directly related to the Army's operational concept (ADP 1-01). Tenets of unified land operations describe the Army's approach to generating and applying combat power across the range of military operations. Commanders and staffs consider and incorporate the following tenets into all plans:
- Simultaneity.
- Depth.
- Synchronization.
- Flexibility.

Simultaneity

2-112. *Simultaneity* is the execution of related and mutually supporting tasks at the same time across multiple locations and domains (ADP 3-0). Army forces employing capabilities simultaneously across the air, land, maritime, space, and cyberspace domains present dilemmas to adversaries and enemies, while reassuring allies and influencing neutrals. Planners consider the simultaneous application of joint and combined arms capabilities across the range of military operations to overwhelm threats physically and psychologically.

2-113. Simultaneous operations across multiple domains—conducted in depth and supported by military deception—present the enemy with multiple dilemmas. These operations degrade enemy freedom of action, reduce enemy flexibility and endurance, and disrupt enemy plans and coordination. Such operations place critical enemy functions at risk and deny the enemy the ability to synchronize or generate combat power. The application of capabilities in a complementary and reinforcing fashion creates more problems than the enemy commander can solve, which erodes both enemy effectiveness and the will to fight.

Depth

2-114. *Depth* is the extension of operations in time, space, or purpose to achieve definitive results (ADP 3-0). Commanders use depth to obtain space for effective maneuver, time to conduct operations, and resources to achieve and exploit success. Planners develop ways for forces to engage the enemy throughout their depth, preventing the effective employment of reserves, and disrupting command and control, logistics, and other capabilities not in direct contact with friendly forces. Operations in depth can disrupt the enemy's decision cycle. They contribute to protection by destroying enemy capabilities before the enemy can use them. In operations, staying power—depth of action—comes from adequate resources. Depth of resources in quantity, positioning, and mobility is critical to executing operations.

Chapter 2

Synchronization

2-115. *Synchronization* is the arrangement of military actions in time, space, and purpose to produce maximum relative combat power at a decisive place and time (JP 2-0). Synchronization is not the same as simultaneity; it is the ability to execute multiple related and mutually supporting tasks in different locations at the same time. These actions produce greater effects than executing each in isolation. For example, synchronizing information collection, obstacles, direct fires, and indirect fires results in destroying an enemy formation during a defense. When conducting offensive tasks, synchronizing forces along multiple lines of operations temporarily disrupts the enemy organization and creates opportunities for an exploitation.

2-116. Commanders determine the degree of control necessary to synchronize their operations. They balance synchronization with agility and initiative, never surrendering the initiative for the sake of synchronization. Excessive synchronization can lead to too much control, which limits the initiative of subordinates and undermines flexibility.

Flexibility

2-117. *Flexibility* is the employment of a versatile mix of capabilities, formations, and equipment for conducting operations (ADP 3-0). To achieve tactical, operational, and strategic success, effective commanders adapt to conditions as they change and employ forces in multiple ways. Flexibility facilitates collaborative planning and decentralized execution. Leaders learn from experience (their own and that of others) and apply new knowledge to each situation. Flexible plans help units adapt quickly to changing circumstances during operations. Flexible plans provide options to commanders for addressing new or unforeseen circumstances during execution. Ultimately, flexibility enables commanders to mitigate risk.

Tenets in Action: OPERATION JUST CAUSE

Late on 19 December 1989, a joint force of 7,000 Soldiers, sailors, airmen, and marines deployed from U.S. bases bound for Panama. During the early morning hours of 20 December, this force—supported by the United States Southern Command (USSOUTHCOM) forward-deployed forces in Panama—simultaneously hit targets at 26 separate locations across the depth of the country.

The success of the attack against key Panamanian Defense Force (PDF) strongholds required the synchronization of multiple actions by an assortment of U.S. special operations forces and elements from the 82d Airborne Division, the 5th Mechanized Division, the 7th Infantry Divisions, and Marine Corps. These were supported by the Air Force and Navy in various ways, including airlift and sealift, suppression of enemy air defense, and AC-130 gunship strikes. Subordinate initiative during execution contributed greatly to the ability of the joint force to rapidly paralyze PDF response capability.

COMMANDERS FOCUS PLANNING

2-118. The responsibility for planning is inherent in command. Commanders are planners—they are the central figure to effective planning. Often with the most experience, commanders are ultimately responsible for the execution of the plan. As such, the plan must reflect how commanders intend to conduct operations. Commanders ensure the approaches to planning meet the requirements of time, planning horizons, level of detail, and desired outcomes. Commanders ensure that all plans and orders comply with domestic and international laws as well as the Army Ethic. They confirm that the plan or order is relevant and suitable for subordinates. Generally, the more involved commanders are in planning, the faster staffs can plan. Through personal involvement, commanders learn from the staff and others about a situation and ensure the plan reflects their commander's intent.

DEVELOP SIMPLE, FLEXIBLE PLANS THROUGH MISSION ORDERS

It is my opinion that Army orders should not exceed a page and a half of typewritten text and it was my practice not to issue orders longer than this. Usually they can be done on one page, and the back of the page used for a sketch map.

General George S. Patton, Jr.

2-119. Simplicity—prepare clear, uncomplicated plans and clear, concise orders to ensure thorough understanding—is a principle of war. Effective plans and orders are simple and direct. Staffs prepare clear, concise orders that communicate understanding of the operation by using doctrinally correct military terms and symbols. Using the correct terms and symbols minimizes chances of misunderstanding and aids with brevity. Developing shorter plans helps maintain simplicity. Shorter plans are easier to disseminate, read, and remember.

2-120. Complex plans requiring intricate coordination or having inflexible timelines have a greater potential to fail during execution. Operations are always subject to friction beyond the control of commanders and staffs. Elaborate or complex plans that do not incorporate tolerances for friction have more chances of something irrevocable going wrong. Simple plans are more responsive to changes in enemy behavior, the weather, and issues with friendly forces.

2-121. Simple plans require an easily understood concept of operations. Planners promote simplicity by minimizing details where possible and by limiting the actions or tasks to what the situation requires. Subordinates can then develop specifics within the commander's intent. For example, instead of assigning a direction of attack, planners can designate an axis of advance.

2-122. Simple plans are not simplistic plans. Simplistic refers to something made overly simple by ignoring the situation's complexity. Good plans simplify complicated situations. However, some situations require more complex plans than others do. Commanders at all levels weigh the apparent benefits of a complex concept of operations against the risk that subordinates will be unable to understand or follow it. Commanders prefer simple plans because units can understand and execute them more easily.

2-123. Flexible plans help units adapt quickly to changing circumstances. Commanders and planners build opportunities for initiative into plans by anticipating events. This allows them to operate inside of the enemy's decision cycle or to react promptly to deteriorating situations. Incorporating options to reduce risk adds flexibility to a plan. Identifying decision points and designing branches and sequels ahead of time—combined with a clear commander's intent—helps create flexible plans.

2-124. Commanders stress the importance of using mission orders as a way of building simple, flexible plans. *Mission orders* are directives that emphasize to subordinates the results to be attained, not how they are to achieve them (ADP 6-0). Mission orders are not a specific type of order but a reflected style or technique for writing OPLANs, OPORDs, and FRAGORDs. In developing mission orders, commanders focus subordinates on what to do and why to do it without prescribing exactly how to do it. Commanders establish control measures to aid cooperation among forces without imposing needless restrictions on freedom of action.

2-125. Mission orders clearly convey the unit's mission and commander's intent. They summarize the situation, describe the operation's objectives and end state, and provide a simple concept of operations to accomplish the mission. When assigning tasks to subordinate units, mission orders include all components of a task statement: who, what, when, where, and why. However, a task statement emphasizes the purpose (why) of the tasks to guide (along with the commander's intent) subordinate initiative.

2-126. Mission orders contain the proper level of detail; they are neither so detailed that they stifle initiative nor so general that they provide insufficient direction. The proper level of detail is situationally dependent. Some phases of operations require tighter control over subordinate elements than others require. An air assault's air movement and landing phases, for example, require precise synchronization. Its ground maneuver plan requires less detail. As a rule, the base plan or order contains only the specific information required to provide the guidance to synchronize combat power at the decisive time and place while allowing subordinates as much freedom of action as possible. Commanders rely on subordinate initiative and coordination to act within the commander's intent and concept of operations.

Chapter 2

OPTIMIZE AVAILABLE PLANNING TIME

You can ask me for anything you like, except for time.

Napoleon Bonaparte

2-127. Time is a critical variable in all operations. Therefore, time management is important in planning. Whether done deliberately or rapidly, all planning requires the skillful use of available time to optimize planning and preparation throughout the unit. Taking more time to plan often results in greater synchronization; however, any delay in execution risks yielding the initiative—with more time to prepare and act—to the enemy.

2-128. When allocating planning time to staffs, commanders ensure subordinates have enough time to plan and prepare their own actions prior to execution. Commanders follow the "one-third, two-thirds rule" as a guide to allocate time available. They use one-third of the time available before execution for their own planning and allocate the remaining two-thirds of the time available before execution to their subordinates for planning and preparation.

2-129. Both collaborative planning and parallel planning help optimize available planning time. **Collaborative planning is two or more echelons planning together in real time, sharing information, perceptions, and ideas to develop their respective plans simultaneously.** This type of planning speeds planning efforts as organizations share their understanding of the situation, participate in COA development and decision making together, and develop their respective plans or orders as opposed to waiting for a higher echelon to complete the plan prior to beginning planning.

2-130. ***Parallel planning* is two or more echelons planning for the same operations nearly simultaneously facilitated by the use of warning orders by the higher headquarters.** In this type of planning, several echelons developing their plans in parallel significantly shorten planning time across the force. The higher headquarters shares information concerning future operations with subordinate units through WARNORDs and other means. Frequent communication between commanders and staffs and sharing of information (such as IPB products) help subordinate headquarters plan. Parallel planning is used when time is of the essence and the likelihood of execution of the plan is high.

2-131. Commanders are careful not to burden subordinates with planning requirements too far into the future, instead enabling subordinates to focus on execution. Generally, the higher the headquarters, the more time and resources staff have available to plan and explore options. Higher headquarters involve subordinates with developing those plans and concepts that have the highest likelihood of being adopted.

FOCUS ON THE RIGHT PLANNING HORIZON

The process of preparing combat orders varies widely to the situation. Days or weeks may be devoted to the task by the commander and his staff. On the other hand, instant action may be called for especially in the division and lower units. The commander and his staff must be able to adapt their procedure to any situation encountered.

FM 101-5, *Staff Officers Field Manual* (1940)

2-132. The defining challenges to effective planning are uncertainty and time. Tension exists when commanders determine how far ahead to plan effectively without preparation and coordination becoming irrelevant. Planning too far into the future may overwhelm the capabilities of planning staffs, especially subordinate staffs. Not planning far enough ahead may result in losing the initiative and being unprepared. Understanding this tension is key to ensuring that the command focuses on the right planning horizon.

2-133. **A *planning horizon* is a point in time commanders use to focus the organization's planning efforts to shape future events.** Planning horizons may be measured in weeks or months or in hours and days depending on the echelon and situation. Organizations often plan simultaneously in several different horizons, especially division and above. To guide their planning efforts, commanders use three planning horizons—short-range, mid-range, and long-range.

2-134. The range of planning directly correlates with the certainty commanders have of attaining the end state. Short-range planning is conducted under conditions of relative certainty when commanders believe

Planning

they can reasonably forecast events, assign resources, and commit to a particular plan. Short-range planning normally results in an OPORD or FRAGORD for execution. In conditions of moderate certainty, mid-range planning focuses on developing several options to the base plan normally resulting in a branch plan or sequel. Beyond the mid-term planning horizon, the situation is often too uncertain to develop detailed operational plans. Instead, commanders develop broad concepts (for example, an OPLAN in concept form) addressing a number of different circumstances over a longer period. These plans vary in level of detail based on assumptions about the future that address "what if?" scenarios.

Note. In addition to planning horizons, commanders also prioritize planning efforts using conceptual focus areas such as a line of effort, a specific objective, or problem set.

DETERMINE RELEVANT FACTS AND DEVELOP ASSUMPTIONS

Since all information and assumptions are open to doubt, and with chance at work everywhere, the commander continually finds that things are not as he expected. This is bound to influence his plans, or at least the assumptions underlying them.

Carl von Clausewitz

2-135. Commanders and staffs gather key facts and develop assumptions as they build their plan. A fact is something known to exist or have happened—a statement known to be true. Facts concerning the operational and mission variables serve as the basis for developing situational understanding during planning. When listing facts, planners are careful they are directly relevant to a COA or help commanders make a decision. Any captured, recorded, and most importantly briefed fact must add value to the planning conversation.

2-136. An assumption provides a supposition about the current situation or future course of events, presumed to be true in the absence of facts. Assumptions must be valid (logical and realistic) and necessary for planning to continue. Assumptions address gaps in knowledge that are critical for the planning process to continue. Staffs continually review assumptions to ensure validity and to challenge if they appear unrealistic.

2-137. Commanders and staffs use care with assumptions to ensure they are not based on preconceptions; bias; false historical analogies; or simple, wishful thinking. Additionally, effective planners recognize any unstated assumptions. Accepting a broad assumption without understanding its sublevel components often leads to other faulty assumptions. For example, a division commander might assume a combined arms battalion from the continental United States is available in 30 days. This commander must also understand the sublevel components—adequate preparation, load and travel time, viable ports and airfields, favorable weather, and enemy encumbrance. The commander considers how the sublevel components hinder or aid the battalion's ability to be available.

2-138. Commanders and staffs continuously question whether their assumptions are valid throughout planning and the operations process. Key points concerning the use of assumptions include—

- Assumptions must be logical, realistic, and considered likely to be true.
- Assumptions are necessary for continued planning.
- Too many assumptions result in a higher probability that the plan or proposed solution may be invalid.
- The use of assumptions requires the staff to develop branches to execute if one or more key assumptions prove false.
- Often, an unstated assumption may prove more dangerous than a stated assumption proven wrong.

PLANNING PITFALLS

In war, leaders of small units are usually no more than one or two jumps ahead of physical and mental exhaustion. In addition, they run a never-ending race against time. In such conditions long, highly involved orders multiply the ever-present chance of misunderstanding, misinterpretation, and plain oversight.

Infantry in Battle (1939)

2-139. Commanders and staffs recognize the value of planning and avoid common planning pitfalls. These pitfalls generally stem from a common cause: the failure to appreciate the unpredictability and uncertainty of military operations. Pointing these out is not a criticism of planning, but of planning improperly. Common planning pitfalls include—

- Attempting to forecast and dictate events too far into the future.
- Trying to plan in too much detail.
- Using the plan as a script for execution.
- Institutionalizing rigid planning methods.

2-140. The first pitfall, attempting to forecast and dictate events too far into the future, may result from believing a plan can control the future. Planners tend to plan based on assumptions that the future will be a linear continuation of the present. These plans often underestimate the scope of changes in directions that may occur and the results of second- and third-order effects. Even the most effective plans cannot anticipate all the unexpected events. Often, events overcome plans much sooner than anticipated. Effective plans include sufficient branches and sequels to account for the nonlinear nature of events.

2-141. The second pitfall consists of trying to plan in too much detail. Sound plans include necessary details; however, planning in unnecessary detail consumes limited time and resources that subordinates need. This pitfall often stems from the desire to leave as little as possible to chance. In general, the less certain the situation, the fewer details a plan should include. However, planners often respond to uncertainty by planning in more detail to try to account for every possibility. Preparing detailed plans under uncertain conditions generates even more anxiety, which leads to even more detailed planning. Often this over planning results in an extremely detailed plan that does not survive the friction of the situation and constricts effective action. A good plan only includes details needed to coordinate or synchronize actions of two or more subordinate units.

2-142. The third pitfall, using the plan as a script for execution, tries to prescribe the course of events with precision. When planners fail to recognize the limits of foresight and control, the plan can become a coercive and overly regulatory mechanism. Commanders, staffs, and subordinates mistakenly focus on meeting the requirements of the plan rather than deciding and acting effectively.

2-143. The fourth pitfall is the danger of institutionalizing rigid planning methods that leads to inflexible or overly structured thinking. This pitfall tends to make planning rigidly focused on the process and produces plans that overly emphasize detailed procedures. Effective planning provides a disciplined framework for approaching and solving complex problems. Taking that discipline to the extreme often results in subordinates not getting plans on time or getting overly detailed plans.

Chapter 3
Preparation

The stroke of genius that turns the fate of a battle? I don't believe in it. A battle is a complicated operation that you prepare laboriously.

Marshal Ferdinand Foch

This chapter addresses the fundamentals of preparations to include its definition and functions. It offers guidelines for effective preparation and addresses specific preparation activities commonly performed within the headquarters and across the force to improve the unit's ability to execute operations.

FUNDAMENTALS OF PREPARATION

3-1. ***Preparation* consists of those activities performed by units and Soldiers to improve their ability to execute an operation.** Preparation creates conditions that improve friendly forces' opportunities for success and include activities such as rehearsals, training, and inspections. It requires commander, staff, unit, and Soldier actions to ensure the force is ready to execute operations.

3-2. Preparation helps the force transition from planning to execution. Preparation normally begins during planning and continues into execution by uncommitted units. Like the other activities of the operations process, commanders drive preparation activities with a focus on leading and assessing. The functions of preparation include the following:
- Improve situational understanding.
- Develop a common understanding of the plan.
- Train and become proficient on critical tasks.
- Task-organize and integrate the force.
- Ensure forces and resources are positioned.

IMPROVE SITUATIONAL UNDERSTANDING

3-3. Developing and maintaining situational understanding requires continuous effort throughout the operations process as discussed in chapter 1. During preparation, commanders continue to improve their understanding of a situation. They realize that the initial understanding developed during planning may be neither accurate nor complete. As such, commanders strive to validate assumptions and improve their situational understanding as they prepare for operations. Information collection helps leaders better understand the enemy, terrain, and civil considerations. Inspections, rehearsals, and liaison help leaders improve their understanding of the friendly force. Based on new information gained from various preparation activities, commanders refine the plan prior to execution.

DEVELOP A COMMON UNDERSTANDING OF THE PLAN

Before the battle begins an Army Commander should assemble all commanders down to the lieutenant-colonel level and explain to them the problem, his intention, his plan, and generally how he is going to fight the battle and make it go the way he wants....

Field Marshal Bernard Law Montgomery

3-4. A successful transition from planning to execution requires those charged with executing the order to understand the plan fully. The transition between planning and execution takes place both internally in the headquarters and externally between the commander and subordinate commanders. Several preparation activities help leaders develop a common understanding of the plan. Confirmation briefs, rehearsals, and the plans-to-operations transition brief help improve understanding of the concept of operations, control measures, decision points, and command and support relationships. They assist the force with understanding the plan prior to execution.

TRAIN AND BECOME PROFICIENT ON CRITICAL TASKS

3-5. Units train to become proficient in those tasks critical to success for a specific operation. Commanders issue guidance about which tasks to train on and rehearse. They emphasize training that incorporates attached units to ensure they are integrated and to mitigate interoperability challenges prior to execution. Commanders also allocate time during preparation for units and Soldiers to train on unfamiliar tasks prior to execution. For example, a unit unfamiliar with a wet-gap crossing requires significant training and familiarization on employing small boats. Units may need to train on crowd control techniques in support of host-nation elections. Leaders also allocate time for attaining and maintaining proficiency on individual Soldier skills (such as zeroing weapons, combat lifesaving tasks, and language familiarization).

TASK-ORGANIZE AND INTEGRATE THE FORCE

3-6. Task-organizing the force is an important part of planning. During preparation, commanders allocate time to put the new task organization into effect. This includes detaching units, moving forces, and receiving and integrating new units and Soldiers into the force. When units change task organization, they need preparation time to learn the gaining unit's SOPs and learn their role in the gaining unit's plan. The gaining unit needs preparation time to assess the new unit's capabilities and limitations and to integrate new capabilities. Properly integrating units and Soldiers into the force builds trust and improves performance in execution.

ENSURE FORCES AND RESOURCES ARE POSITIONED

3-7. Effective preparation ensures the right forces are in the right place at the right time with the right equipment and other resources ready to execute the operation. Concurrent with task organization, commanders use troop movement to position or reposition forces to the correct locations prior to execution. This includes positioning sustainment units and supplies. Pre-operations checks confirm that the force has the proper and functional equipment before execution.

GUIDES TO EFFECTIVE PREPARATION

> *Nine-tenths of tactics are certain, and taught in books; but the irrational tenth is like the kingfisher flashing across the pool, and that is the test of generals. It can only be ensured by instinct, sharpened by thought practicing the stroke so often that at the crisis it is as natural as a reflex.*
>
> T. E. Lawrence

3-8. Like the other activities of the operations process, commanders drive preparation. They continue to understand, visualize, describe, direct, lead, and assess. They gather additional information to improve their situational understanding, revise the plan as required, coordinate with other units and partners, and supervise preparation activities to ensure their forces are ready to execute operations. The following guides aid commanders and leaders in effectively preparing for operations:

- Allocate time and prioritize preparation efforts.
- Protect the force.
- Supervise.

Preparation

> ### Prepare: Rangers Train for Seizing Pointe du Hoc
>
> The Allied plan for the 1944 invasion of the Normandy coast was divided into five significant landing sights: Juno, Sword, Gold, Utah, and Omaha Beaches. The Canadians would hit Juno Beach while the British stormed Sword and Gold Beaches. The Americans would strike Utah and Omaha Beaches as well as the high cliff between these beaches called Pointe du Hoc.
>
> It was on Pointe du Hoc that the Germans had constructed a heavily fortified coastal artillery position. Of note, it had six 155 mm guns facing the English Channel. These big guns enabled a secure defense, as the guns could rain havoc on the invasion force on Omaha Beach and on the ships in the channel. To protect this significant position, the Germans built cement gun encasements, designed an interlocking trench system, including underground trenches, and deployed anti-aircraft artillery. Pointe du Hoc appeared impregnable.
>
> The most unexpected route of attack was from the sea. The Americans, however, considered it an accessible assault point. They reasoned that with a well-trained force at low tide, Soldiers could land on the narrow beaches below and ascend the cliff with the assistance of ropes and ladders. Understanding the hazards and vital importance of the landing beaches along the coast, Supreme Allied Commander General Dwight D. Eisenhower assigned the mission for the assault on Pointe du Hoc to Lieutenant Colonel James E. Rudder and elements of the 2d and 5th Ranger Battalions.
>
> The Rangers spent considerable time learning new skills and rehearsing for what many considered a suicide mission. While the Rangers received some instruction from British commandos, the Rangers mostly learned cliff climbing by trial and error. The Rangers practiced with various types of ropes and ladders. Eventually rocket-fired rappelling ropes equipped with grappling hooks became the primary tool of choice for ascending the cliff of Pointe du Hoc. Specially designed landing crafts outfitted with rocket launchers deployed the ropes to the tops of the cliff, 100 feet up from the beach. The Rangers also added extension ladders to several of the larger landing craft to extend towards the cliff top. They mounted machine guns on the top rungs of these ladders to suppress German machine gunners.
>
> In the weeks leading up to D-Day, the Rangers trained, developed, and tested their newly formed skills on various cliffs along the English coast and on the Isle of Wight.
>
> Eisenhower, at 0330 hours on June 5, decided that the favorable weather greenlighted OPERATION OVERLORD by shouting "Ok let's go." Shortly before 0400 on June 6, the Rangers manned their landing craft fully prepared for their mission to destroy the German guns at Pointe du Hoc.

ALLOCATE TIME AND PRIORITIZE PREPARATION EFFORTS

3-9. Mission success depends as much on preparation as on planning. Higher headquarters may develop the best of plans; however, plans serve little purpose if subordinates do not receive them in time to understand them, develop their own plans, and prepare for the upcoming operation. As part of the operational timeline, commanders allocate sufficient time for preparation. This includes time for detaching units, moving forces, and receiving and integrating new units and Soldiers into the force. It includes time to rehearse the operation to include designating the type of rehearsals. Commanders prioritize preparation activities by issuing specific instructions in WARNORDs and the OPORD.

Chapter 3

PROTECT THE FORCE

3-10. The force as a whole is often most vulnerable to surprise and enemy attack during preparation. As such, security operations—screen, guard, cover, area security, and local security—are essential during preparation. In addition, commanders ensure integration of the various tasks of the protection warfighting function to safeguard bases, secure routes, recover isolated personnel, and protect the force while it prepares for operations. (See ADP 3-37 for a detailed discussion of protection.) Operations security is an important consideration during preparation. Commanders direct measures to reduce the vulnerabilities of friendly action to enemy observation and exploitation. This includes measures to hide friendly movements, rehearsals, and the movement and concentration of forces.

SUPERVISE

3-11. Attention to detail is critical to effective preparation. Leaders monitor and supervise activities to ensure the unit is ready for the mission. Leaders supervise subordinates and inspect their personnel and equipment. Rehearsals allow leaders to assess their subordinates' preparations. They may identify areas that require more supervision.

PREPARATION ACTIVITIES

If I always appear prepared, it is because before entering on an undertaking, I have meditated for long and have foreseen what may occur. It is not genius which reveals to me suddenly and secretly what I should do in circumstances unexpected by others; it is thought and preparation.

Napoleon Bonaparte

3-12. Preparation activities help commanders, staffs, and Soldiers understand a situation and their roles in upcoming operations. Commanders, units, and Soldiers conduct the activities listed in table 3-1 to help ensure the force is prepared for execution.

Table 3-1. Preparation activities

• Coordinate and establish liaison	• Initiate sustainment preparation
• Initiate information collection	• Initiate network preparations
• Initiate security operations	• Manage terrain
• Initiate troop movements	• Prepare terrain
• Complete task organization	• Conduct confirmation briefs
• Integrate new units and Soldiers	• Conduct rehearsals
• Train	• Conduct plans-to-operations transition
• Conduct pre-operations checks and inspections	• Revise and refine the plan
	• Supervise

COORDINATE AND ESTABLISH LIAISON

3-13. Units and organizations establish liaison in planning and preparation. Establishing liaison helps leaders internal and external to the headquarters understand their unit's role in upcoming operations and prepare to perform that role. In addition to military forces, many civilian organizations may operate in the operational area. Their presence can both affect and be affected by the commander's operations. Continuous liaison between the command and unified action partners helps to build unity of effort.

3-14. Liaison is most commonly used for establishing and maintaining close communications. It continuously enables direct, physical communications between commands. Establishing and maintaining liaison is vital to external coordination. Liaison enables direct communications between the sending and receiving headquarters. It may begin with planning and continue through preparing and executing, or it may start as late as execution. Available resources and the need for direct contact between sending and receiving headquarters determine when to establish liaison. (See FM 6-0 for a detailed discussion of liaison.)

3-15. Establishing liaisons with civilian organizations is especially important in stability operations because of various external organizations and the inherent coordination challenges. Civil affairs units (to include LNOs) are particularly important in coordination with civilian organizations.

INITIATE INFORMATION COLLECTION

3-16. During planning and preparation, commanders take every opportunity to improve their situational understanding prior to execution. This requires aggressive and continuous information collection. Commanders often direct information collection (to include reconnaissance operations) early in planning that continues in preparation and execution. Through information collection, commanders and staffs continuously plan, task, and employ collection assets and forces to collect timely and accurate information to help satisfy CCIRs and other information requirements. (See FM 3-55 for discussion on information collection.)

INITIATE SECURITY OPERATIONS

3-17. Security operations—screen, guard, cover, area security, and local security—are essential during preparation. During preparation, the force is vulnerable to surprise and enemy attacks. Leaders are often away from their units and concentrated together during rehearsals. Parts of the force could be moving to task-organize. Required supplies may be unavailable or being repositioned. Units assigned security missions execute these missions while the rest of the force prepares for the overall operation. Every unit provides local security to its own forces and resources.

INITIATE TROOP MOVEMENTS

3-18. The repositioning of forces prior to execution makes up a significant portion of activities of preparation. Commanders position or reposition units to the correct starting places before execution. Commanders integrate operations security measures with troop movements to ensure these movements do not reveal any intentions to the enemy. Troop movements include assembly area reconnaissance by advance parties and route reconnaissance. They also include movements required by changes to the task organization. Commanders can use WARNORDs to direct troop movements before they issue the OPORD.

COMPLETE TASK ORGANIZATION

3-19. During preparation, commanders complete task-organizing their force to obtain the right mix of capabilities to accomplish a specific mission. The commander may direct task organization to occur immediately before the OPORD is issued. This task-organizing is done with a WARNORD. Doing this gives units more time to execute the tasks needed to affect the new task organization. Task-organizing early allows affected units to become better integrated and more familiar with all elements involved. This is especially important with inherently time-consuming tasks, such as planning technical network support for the organization.

INTEGRATE NEW UNITS AND SOLDIERS

3-20. Commanders, command sergeants major, and staffs help assimilate new units into the force and new Soldiers into their units. They also prepare new units and Soldiers in performing their duties properly and integrating into an upcoming operation smoothly. Integration for new Soldiers includes training on unit SOPs and mission-essential tasks for the operation. It also means orienting new Soldiers on their places and roles in the force and during the operation. This integration for units includes, but is not limited to—

- Receiving and introducing new units to the force and the AO.
- Exchanging SOPs.
- Conducting briefs and rehearsals.
- Establishing communications links.
- Exchanging liaison teams (if required).

Chapter 3

TRAIN

In no other profession are the penalties for employing untrained personnel so appalling or so irrevocable as in the military.

General Douglas MacArthur

3-21. Training prepares forces and Soldiers to conduct operations according to doctrine, SOPs, and the unit's mission. Training develops the teamwork, trust, and mutual understanding that commanders need to exercise mission command and that forces need to achieve unity of effort. Training does not stop when a unit deploys. If the unit is not conducting operations or recovering from operations, it is training. While deployed, unit training focuses on fundamental skills, current SOPs, and skills for a specific mission.

CONDUCT PRE-OPERATIONS CHECKS AND INSPECTIONS

3-22. Unit preparation includes completing pre-operations checks and inspections. These checks ensure units, Soldiers, and systems are as fully capable and ready to execute the mission as time and resources permit. The inspections ensure the force has the resources necessary to accomplish the mission. During pre-operations checks and inspections, leaders also check Soldiers' ability to perform crew drills that may not be directly related to the mission. Some examples of these include drills that respond to a vehicle rollover or an onboard fire.

INITIATE SUSTAINMENT PREPARATION

3-23. Resupplying, maintaining, and issuing supplies or equipment are major activities during preparation. Repositioning of sustainment assets can also occur. During preparation, sustainment personnel at all levels take action to optimize means (force structure and resources) for supporting the commander's plan. These actions include, but are not limited to, identification and preparation of bases, coordinating for host-nation support, and improving lines of communications.

INITIATE NETWORK PREPARATIONS

3-24. During preparation, units must tailor the information network to meet the specific needs of each operation. This includes not only the communications, but also how the commander expects information to move between and be available for units and leaders in an AO. Commanders and staffs prepare and rehearse the information network to support the plan by—
- Managing available bandwidth.
- Providing availability and location of data and information.
- Positioning and structuring network assets.
- Tracking status of key network systems.

MANAGE TERRAIN

3-25. Terrain management includes allocating terrain by establishing AOs, designating assembly areas, and specifying locations for units. Terrain management is an important activity during preparation as units reposition and stage prior to execution. Commanders assigned an AO manage terrain within their boundaries. Through terrain management, commanders identify and locate units in the area. The operations officer, with support from others in the staff, can then de-conflict operations, control movements, and deter fratricide as units get in position to execute planned missions. Commanders also consider unified action partners located in their AO and coordinate with them for the use of terrain.

PREPARE TERRAIN

3-26. Commanders must understand the terrain and the infrastructure of their AO as early as possible to identify potential for improvement, establish priorities of work, and begin preparing the area. Terrain preparation involves shaping the terrain to gain an advantage, such as building fighting and protective positions, improving cover and concealment, and reinforcing obstacles. Engineer units are critical in assisting units in preparing terrain to include building and maintaining roads, trails, airfields, and bases camps prior to and during operations. (See FM 3-34 for a detailed discussion of engineer operations.)

CONDUCT CONFIRMATION BRIEFS

3-27. A *confirmation brief* **is a brief subordinate leaders give to the higher commander immediately after the operation order is given to confirm understanding.** It is their understanding of the higher commander's intent, their specific tasks, and the relationship between their mission and the other units' missions in the operation. The confirmation brief is a tool used to ensure subordinate leaders understand—

- The commander's intent, mission, and concept of operations.
- Their unit's tasks and associated purposes.
- The relationship between their unit's mission and those of other units in the operation.

Ideally, the commander conducts confirmation briefs in person with selected staff members of the higher headquarters present.

CONDUCT REHEARSALS

> *Sand-Table Exercises by staffs up to and including corps or army, even on the most rudimentary type of sand table, are extremely helpful prior to an attack.*
>
> General George S. Patton, Jr.

3-28. A *rehearsal* **is a session in which the commander and staff or unit practices expected actions to improve performance during execution.** Commanders use rehearsals to ensure staffs and subordinates understand the concept of operations and commander's intent. Rehearsals also allow leaders to practice synchronizing operations at times and places critical to mission accomplishment. Effective rehearsals imprint a mental picture of the sequence of the operation's key actions and improve mutual understanding among subordinate and supporting units and leaders. The extent of rehearsals depends on available time. In cases of short-notice requirements, detailed rehearsals may not be possible. In these instances, subordinate leaders backbrief their higher commander on how they intend to accomplish the mission. (See FM 6-0 for a discussion of the different types of rehearsals.) Leaders conduct rehearsals to—

- Practice essential tasks.
- Identify weaknesses or problems in the plan.
- Coordinate subordinate element actions.
- Improve Soldier understanding of the concept of operations.
- Foster confidence among Soldiers.

CONDUCT PLANS-TO-OPERATIONS TRANSITION

3-29. The plans-to-operations transition is a preparation activity that occurs within the headquarters. It ensures members of the current operations integration cell fully understand the plan before execution. During preparation, the responsibility for maintaining the plan shifts from the plans (or future operations integrating cell for division and above headquarters) integrating cell to the current operations integration cell (see figure 3-1). This transition is the point at which the current operations integration cell becomes responsible for short-term planning and controlling execution of the OPORD. This responsibility includes answering requests for information concerning the order and maintaining the order through FRAGORDs. This transition enables the plans and future operations integrating cells to focus their planning efforts on sequels, branches, and other planning requirements directed by the commander.

Chapter 3

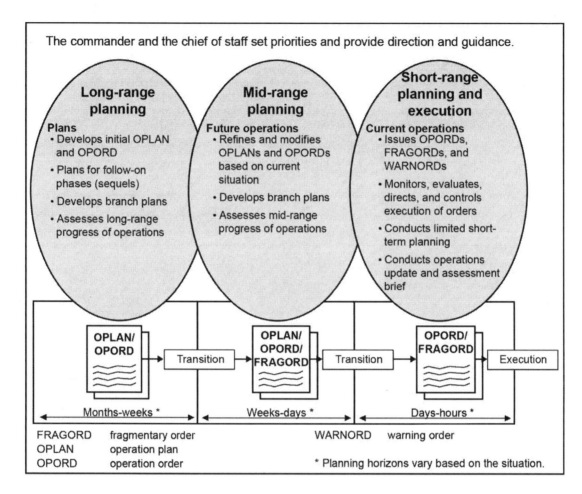

Figure 3-1. Transition among the integrating cells

3-30. The timing of the plans-to-operations transition requires careful consideration. It must allow enough time for members of the current operations integration cell to understand the plan well enough to coordinate and synchronize its execution. Ideally, the plans or future operations cell briefs members of the current operations integration cell on the transition before the combined arms rehearsal. This brief enables members of the current operations integration cell to understand the upcoming operation as well as identify friction points and issues to solve prior to its execution. The transition brief is a mission brief that generally follows the five-paragraph OPORD format. Additional areas addressed, include—
- Decision support products (execution matrixes, decision support templates, decision support matrixes, and risk assessment matrixes).
- Known friction points.
- Branches and sequels under considerations.
- Outstanding requests for information and issues.

3-31. Following the combined arms rehearsal, planners and members of the current operations integration cell review additional planning guidance issued by the commander and modify the plan as necessary. Significant changes may require assistance from the plans cell to include moving a lead planner to the current operations integration cell. The plans cell continues planning for branches and sequels.

REVISE AND REFINE THE PLAN

3-32. Revising and refining the plan is a key activity of preparation. During preparation, assumptions made during planning may be proven true or false. Intelligence analysis and reconnaissance may confirm or deny enemy actions or show changed conditions in the AO because of shaping operations. The status of friendly forces may change as the situation changes. Rehearsals may identify coordination issues or other problems needing adjusted. In any of these cases, commanders identify the changed conditions, refine the plan, and issue FRAGORDs.

SUPERVISE

3-33. When leaders supervise, they check details critical to effective preparation. Leaders monitor and supervise activities to ensure the unit is ready for the mission. Leaders supervise subordinates and inspect their personnel and equipment. Rehearsals allow leaders to assess their subordinates' preparations. They may identify areas that require more supervision.

Large-Unit Preparation: Third Army Readies for OPERATION IRAQI FREEDOM

At an October 2002 Army commanders' conference, General Eric Shinseki directed, "From today forward the main effort of the US Army must be to prepare for war with Iraq." Third Army, already supporting operations in Afghanistan, shifted emphasis from deterring an Iraqi invasion of Kuwait or Saudi Arabia to mounting offensive operations to topple Saddam Hussein.

As OPERATION IRAQI FREEDOM changed from possible to probable, the Third Army (also designated as the Coalition Forces Land Component Command [CFLCC]) undertook a number of important tasks designed to prepare for war. From the Army's perspective, these included revising plans for the invasion, preparing the theater infrastructure, determining the ground forces command and control architecture, augmenting and training headquarters and Soldiers, fielding new equipment, providing theater-wide support, mobilizing the Army Reserve and Army National Guard forces, deploying forces into theater, and moving to the border.

Forces prepared to operate at a scale and scope not seen since OPERATION DESERT STORM with units not used to working together. They conducted a series of exercises to refine the plan and to develop procedures, teamwork, and familiarity across the divisions, corps, CFLCC, and U.S. Central Command. EXERCISE VICTORY STRIKE enabled the V Corps staff to practice planning, preparing, and executing corps operations with the focus of deep fires and maneuver that would be critical to the coming campaign. EXERCISE LUCKY WARRIOR provided the first opportunity for the CFLCC's major subordinate elements (V Corps, I Marine Expeditionary Force, and coalition forces) to rehearse operations under the CFLCC headquarters. EXERCISE INTERNAL LOOK provided the venue to the final preparations for the anticipated campaign. This venue allowed the joint force and functional components to examine and refine their plans and work out procedures at the combatant command level. V Corps conducted the last significant series of exercises at Grafenwoehr, Germany in January and February of 2003 called EXERCISE VICTORY SCRIMMAGE.

Collective training by units from the smallest sections all the way up to CFLCC headquarters continued through March. Throughout Kuwait, units engaged in last-minute training; resupplied ammunition, fuel, and other supplies; and ensured individual mental and physical preparedness for war by Soldiers. By 18 March, Third Army and its subordinate units were prepared to open the campaign.

This page intentionally left blank.

Chapter 4
Execution

No plan of operations goes with any degree of certainty to beyond the first contact with the hostile main force.

Field Marshal Helmuth von Moltke

This chapter defines, describes, and offers guidelines for effective execution. It describes the role of the commander and staff during execution followed by a discussion of the major activities of execution. It concludes with a discussion of the rapid decision-making and synchronization process.

FUNDAMENTALS OF EXECUTION

4-1. **Execution is the act of putting a plan into action by applying combat power to accomplish the mission and adjusting operations based on changes in the situation.** In execution, commanders, staffs, and subordinate commanders focus their efforts on translating decisions into actions. They direct action to apply combat power at decisive points and times to achieve objectives and accomplish missions. Inherent in execution is deciding whether to execute planned actions (such as phases, branches, and sequels) or to modify the plan based on unforeseen opportunities or threats.

4-2. Commanders fight the enemy, not the plan. Moltke's dictum above, rather than condemning the value of planning, reminds commanders, staffs, and subordinate unit leaders the proper relationship between planning and execution. A plan provides a reasonably forecast of execution. However, it remains a starting point, not an exact script to follow. As General George S. Patton, Jr. cautioned, "…one makes plans to fit circumstances and does not try to create circumstances to fit plans."

4-3. During execution, the situation may change rapidly. Operations the commander envisioned in the plan may bear little resemblance to actual events in execution. Subordinate commanders need maximum latitude to take advantage of situations and meet the higher commander's intent when the original order no longer applies. Effective execution requires leaders trained in independent decision making, aggressiveness, and risk taking in an environment of mission command.

GUIDES TO EFFECTIVE EXECUTION

I am heartily tired of hearing about what Lee is going to do. Some of you always seem to think he is suddenly going to turn a double somersault, and land in our rear and on both of our flanks at the same time. Go back to your command, and try to think what we are going to do ourselves, instead of what Lee is going to do.

Lieutenant General Ulysses S. Grant

4-4. Execution is a concerted effort to seize and retain the initiative, maintain momentum, and exploit success. Initiative is fundamental to success in any operation, yet simply seizing the initiative is not enough. A sudden barrage of precision munitions may surprise and disorganize the enemy, but if not followed by swift and relentless action, the friendly advantage diminishes and disappears. Successful operations maintain the momentum generated by initiative and exploit successes within the commander's intent. Guides to effective execution include—

- Seize and retain the initiative.
- Build and maintain momentum.
- Exploit success.

Chapter 4

SEIZE AND RETAIN THE INITIATIVE

4-5. Operationally, initiative is setting or dictating the terms of action during operations. Army forces do this by forcing the enemy to respond to friendly actions. By presenting the enemy with multiple cross-domain dilemmas, commanders force the enemy to react continuously until the enemy is finally driven into untenable positions. Seizing the initiative pressures enemy commanders into abandoning their preferred options and making costly mistakes. As enemy mistakes occur, friendly forces apply continuous pressure to prevent the enemy from recovering. These actions enable friendly forces to seize opportunities and create new avenues for an exploitation.

4-6. Seizing the initiative ultimately results from forcing an enemy reaction. Commanders identify times and places where they can mass the effects of combat power to relative advantage. To compel a reaction, they threaten something the enemy cares about such as its center of gravity or decisive points leading to it. By forcing the enemy to react, commanders initiate an action-to-reaction sequence that ultimately reduces enemy options to zero. Each action develops the situation further and reduces the number of possibilities to be considered, thereby reducing friendly uncertainty. Each time the enemy must react, its uncertainty increases. Developing the situation by forcing the enemy to react is the essence of seizing the initiative.

4-7. Retaining the initiative involves applying unrelenting pressure on the enemy. Commanders do this by synchronizing the warfighting functions to present enemy commanders with continuously changing combinations of combat power at a tempo they cannot effectively counter. Commanders and staffs use information collection assets to identify enemy attempts to regain the initiative. During execution, commanders create a seamless, uninterrupted series of actions that forces enemies to react immediately and does not allow them to regain synchronization. Ideally, these actions present enemies with multiple dilemmas, the solutions to any one of which increases the enemy's vulnerability to other elements of combat power.

Take Action

I have found again and again that in encounter actions, the day goes to the side that is the first to plaster its opponent with fire. The man who lies low and awaits developments usually comes off second best.

Field Marshal Erwin Rommel

4-8. Commanders and their subordinate leaders create conditions for seizing the initiative with action. Without action, seizing the initiative is impossible. Faced with an uncertain situation, there is a natural tendency to hesitate and gather more information to reduce the uncertainty. Although waiting and gathering information might reduce uncertainty, such inaction will not eliminate it. Waiting may even increase uncertainty by providing the enemy with time to seize the initiative. Effective leaders can manage uncertainty by acting and developing the situation. When the immediate situation is unclear, commanders clarify it by action, not by sitting and gathering information.

Create and Exploit Opportunities

4-9. Events that offer better ways to success are opportunities. Commanders recognize opportunities by continuously monitoring and evaluating the situation. Failure to understand the opportunities inherent in an enemy's action can surrender the initiative. CCIRs must include information requirements that support exploiting opportunities. Commanders encourage subordinates to act within the commander's intent as opportunities occur. Shared understanding of the commander's intent creates an atmosphere conducive to subordinates exercising initiative.

Accept Risk

4-10. Uncertainty and risk are inherent in all military operations. Recognizing and acting on opportunity means taking risks. Reasonably estimating and intentionally accepting risk is not gambling. Carefully determining the risks, analyzing and minimizing as many hazards as possible, and executing a plan that accounts for those hazards contributes to successfully applying military force. Gambling, in contrast, is imprudently staking the success of an entire action on a single, improbable event. Commanders assess risk by answering three questions:

Execution

- Am I minimizing the risk of friendly losses?
- Am I risking the success of the operation?
- Am I minimizing the risk of civilian casualties and collateral damage?

4-11. When commanders embrace opportunity, they accept risk. It is counterproductive to wait for perfect preparation and synchronization. The time taken to fully synchronize forces and warfighting functions in a detailed order could mean a lost opportunity. It is far better to quickly summarize the essentials, get things moving, and send the details later. Leaders optimize the use of time with WARORDs, FRAGORDs, and verbal updates.

4-12. Commanders exercise the art of command when deciding how much risk to accept. As shown in figure 4-1, the commander has several techniques available to reduce the risk associated in a specific operation. Some techniques for reducing risk take resources from the decisive operation, which reduces the concentration of effects at the decisive point. (See ADP 3-90 for a detailed discussion of the art of tactics and risk reduction.)

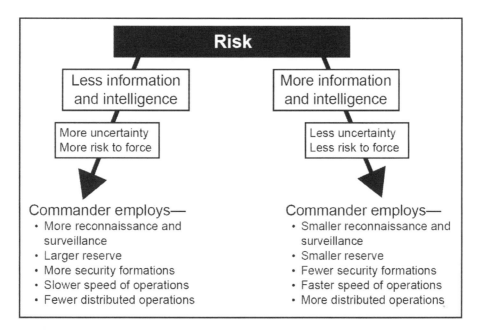

Figure 4-1. Risk reduction factors

BUILD AND MAINTAIN MOMENTUM

4-13. Momentum comes from seizing the initiative and executing decisive, shaping, and sustaining operations at a rapid and sustainable tempo. Momentum allows commanders to create opportunities to engage the enemy from unexpected directions with unanticipated capabilities. Having seized the initiative, commanders continue to control the relative momentum by maintaining focus and pressure and controlling the tempo. They ensure that they maintain momentum by anticipating transitions and moving rapidly between types of operations.

4-14. Speed promotes surprise and can compensate for lack of forces. It magnifies the impact of success in seizing the initiative. By executing at a rapid tempo, Army forces present the enemy with new problems before it can solve current ones. Rapid tempo should not degenerate into haste. Ill-informed and hasty action usually precludes effective combinations of combat power; it may lead to unnecessary casualties.

4-15. The condition of the enemy force dictates the degree of synchronization necessary. When confronted by a coherent and disciplined enemy, commanders may slow the tempo to deliver synchronized attacks. As the enemy force loses cohesion, commanders increase the tempo, seeking to accelerate the enemy's morale and physical collapse.

Chapter 4

EXPLOIT SUCCESS

4-16. Ultimately, only successes that achieve the end state count. To determine how to exploit tactical and operational successes, commanders assess them in terms of the higher commander's intent. However, success will likely occur in ways unanticipated in the plan. Commanders may gain an objective in an unexpected way. Success signals a rapid assessment to answer these questions:

- Does the success generate opportunities that more easily accomplish the objectives?
- Does it suggest other lines of operations or lines of effort?
- Does it cause commanders to change their overall intent?
- Should the force transition to a sequel?
- Should the force accelerate the phasing of the operation?

4-17. An exploitation demands assessment and understanding of the impact on sustaining operations. Sustainment provides the means to exploit success and convert it into decisive results. Sustainment preserves the freedom of action necessary to take advantage of opportunity. Commanders remain fully aware of the status of units and anticipate sustainment requirements; they recognize that sustainment often determines the depth to which Army forces exploit success.

RESPONSIBILITIES DURING EXECUTION

> *To be at the head of a strong column of troops, in the execution of some task that requires brain, is the highest pleasure of war—a grim one and terrible, but which leaves on the mind and memory the strongest mark; to detect the weak point of an enemy's line; to break through with vehemence and thus lead to victory; or to discover some key-point and hold it with tenacity; or to do some other distinct act which is afterward recognized as the real cause of success.*
>
> General William T. Sherman

4-18. During execution, commanders focus their activities on directing, assessing, and leading while improving their understanding and modifying their visualization. Initially, commanders direct the transition from planning to execution as the order is issued and the responsibility for integration passes from the plans cell to the current operations integration cell. During execution, the staff directs units, within delegated authority, to keep the operation progressing successfully. Assessing allows the commander and staff to determine the existence and significance of variances from the operations as envisioned in the initial plan. The staff makes recommendations to the commander about what action to take concerning identified variances in the plan. During execution, leading is as important as decision making, since commanders influence subordinates by providing purpose, direction, and motivation.

COMMANDERS, SECONDS IN COMMAND, AND COMMAND SERGEANTS MAJOR

4-19. During execution, commanders locate where they can best exercise command and sense the operations. Sometimes this is at the command post. Other times, it is forward with a command group. Effective commanders balance the need to make personal observations, provide command presence, and sense the mood of subordinates from forward locations with their ability to maintain command and control continuity with the entire force. No matter where they are located, commanders are always looking beyond the current operation to anticipate the next operation.

4-20. Seconds in command (deputy commanders, executive officers) are a key command resource during execution. First, they can serve as senior advisors to their commander. Second, they may oversee a specific warfighting function (for example, sustainment). Finally, they can command a specific operation (such as a gap crossing), area, or part of the unit (such as the covering force) for the commander.

4-21. The command sergeant major provides another set of senior eyes to assist the commander. The command sergeant major assists the commander with assessing operations as well as assessing the condition and morale of forces. In addition, the command sergeant major provides leadership and expertise to units and Soldiers at critical locations and times during execution.

Staff

4-22. The chief of staff or executive officer is the commander's principal assistant for directing, coordinating, and supervising the staff. During execution, the chief of staff or executive officer must anticipate events and integrate the efforts of the whole staff to ensure operations are proceeding in accordance with the commander's intent and visualization. The chief of staff or executive officer assists the commander with coordinating efforts among the plans, future operations, and current operations integration cells.

4-23. In execution, the staff—primarily through the current operations integration cell—integrates forces and warfighting functions to accomplish the mission. The current operations integration cell is the integrating cell in the command post with primary responsibility for coordinating and directing execution. Staff members in the current operations integration cell actively assist the commander and subordinate units in controlling the current operation. They provide information, synchronize staff and subordinate unit or echelon activities, and coordinate support requests from subordinates. The current operations integration cell solves problems and acts within the authority delegated by the commander. It also performs some short-range planning using the rapid decision-making and synchronization process. (See paragraph 4-36 for a discussion beginning on the rapid decision-making and synchronization process.)

EXECUTION ACTIVITIES

The commander's mission is contained in the orders which he has received. Nevertheless, a commander of a subordinate unit cannot plead absence of orders or the non-receipt of orders as an excuse for inactivity in a situation where action on his part is essential, or where a change in the situation upon which the issued orders were based renders such orders impracticable or impossible of execution.

FM 100-5, *Operations* (1941)

4-24. Execution entails putting the plan into action, and adjusting the plan based on changing circumstances. Friction and uncertainty, especially enemy actions, dynamically affect plans. An accurate situational understanding that accounts for new realties that affect plans provides the basis for commanders to exploit opportunities or counter threats. Major activities of execution include—

- Assessment.
- Decision making.
- Directing action.

Assessment

4-25. During execution, assessment helps commanders visualize probable outcomes and determine whether they need to change the plan to accomplish the mission, take advantage of opportunities, or react to unexpected threats. Assessment includes both monitoring the situation and evaluating progress. Monitoring—the continuous observation of those conditions relevant to the current operation—allows commanders and staffs to improve their understanding of the situation. Evaluation—using indicators to measure change in the situation and judge progress—allows commanders to identify variances, their significance, and if a decision is required to alter the plan. (See chapter 5 for a detailed discussion on assessment.)

4-26. A variance is a difference between the actual situation during an operation and the forecasted plan for the situation at that time or event. A variance can be categorized as an opportunity or threat as shown with the vertical lines in figure 4-2 on page 4-6. The first form of variance is an opportunity to accomplish the mission more effectively. Opportunities result from forecasted or unexpected success. When commanders recognize an opportunity, they alter the order to exploit it if the change achieves the end state more effectively or efficiently. The second form of variance is a threat to mission accomplishment or survival of the force. When recognizing a threat, the commander adjusts the order to eliminate the enemy advantage, restore the friendly advantage, and regain the initiative.

Chapter 4

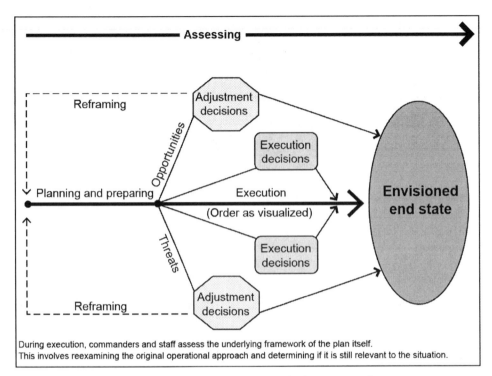

Figure 4-2. Decision making during execution

DECISION MAKING

In war obscurity and confusion are normal. Late, exaggerated or misleading information, surprise situations, and counterorders are to be expected.

Infantry in Battle (1939)

4-27. When operations are progressing satisfactorily, variances are minor and within acceptable levels. Commanders who make this evaluation—explicitly or implicitly—allow operations to continue according to the plan. Plans usually identify some decision points; however, unexpected enemy actions or other changes often present situations that require unanticipated decisions. Commanders act when these decisions are required. As commanders assess the operation, they describe their impressions to staffs and subordinates and then discuss the desirability of choices available. Once commanders make decisions, their staffs transmit the necessary directives, normally in a FRAGORD. Decisions made during execution are either execution decisions or adjustment decisions as shown in figure 4-2's lightly shaded boxes.

Execution Decisions

4-28. Execution decisions implement a planned action under circumstances anticipated in the order such as changing a boundary, committing the reserve, or executing a branch plan. In their most basic form, execution decisions are decisions the commander foresees and identifies for execution during an operation. Commanders are responsible for those decisions but may direct the chief of staff, executive officer, or staff officer to supervise implementation. The current operations integration cell oversees the synchronization and integration needed to implement execution decisions.

Adjustment Decisions

4-29. Adjustment decisions modify the operation to respond to unanticipated opportunities and threats. They often require implementing unanticipated operations and resynchronizing the warfighting functions. Commanders make these decisions, delegating implementing authority only after directing the major change themselves. Adjustments may take one of three forms:

- Reallocating resources.
- Changing the concept of operations.
- Changing the mission.

4-30. The simplest adjustment is reallocating resources. This normally provides additional assets to the decisive operation; however, some situations may require reinforcing a shaping operations. Changing the concept of operations adjusts the way the force executes the operation without changing the mission. Commander's normally do this to exploit an unplanned opportunity or counter an unexpected threat. When reallocating resources or changing the concept of operations does not solve a problem hampering mission accomplishment, the commander may have to change the mission. Commanders change the mission only as a last resort. When they do, the new mission still must accomplish the higher commander' intent. Table 4-1 summarizes a range of possible actions with respect to decisions made during execution.

Table 4-1. Decision types and related actions

	Decision types	Actions
Execution decisions	**Minor variances from the plan** Operation proceeding according to plan. Variances are within acceptable limits.	**To execute planned actions** • Commander or designee decides which planned actions best meet the situation and directs their execution. • Staff issues fragmentary order. • Staff completes follow-up actions.
Execution decisions	**Anticipated situation** Operation encountering variances within the limits for one or more branches or sequels anticipated in the plan.	**To execute a branch or sequel** • Commander or staff reviews branch or sequel plan. • Commander receives assessments and recommendations for modifications to the plan, determines the time available to refine it, and either issues guidance for further actions or directs execution of a branch or sequel. • Staff issues fragmentary order. • Staff completes follow-up actions.
Adjustment decisions	**Unanticipated situation—friendly success** Significant, unanticipated positive variances result in opportunities to achieve the end state in ways that differ significantly from the plan. **Unanticipated situation—enemy threat** Significant, unanticipated negative variances impede mission accomplishment.	**To make an adjustment decision** • Commander recognizes the opportunity or threat and determines time available for decision making. • Commander selects a decision-making method. If there is not enough time for a complete military decision-making process, the commander may direct a single course of action or conduct the rapid decision-making and synchronization process with select staff members. • Depending on time available, commanders may issue verbal fragmentary orders to subordinates followed by a written fragmentary order to counter the threat or exploit an opportunity. • In rare situations, commanders may reframe the problem, change the mission, and develop an entirely new plan to address significant changes in the situation.

Decision-Making Tools

4-31. Several decision support tools assist the commander and staff during execution. Among the most important are the decision support template, decision support matrix, execution matrix, and execution checklist. The current operations integration cell uses these tools, among others, to help control operations and to determine when anticipated decisions are coming up for execution.

4-32. The decision support template depicts decision points, timelines associated with movement of forces and the flow of the operation, and other key items of information required to execute a specific friendly COA. Part of the decision support template is the *decision support matrix*—**a written record of a war-gamed course of action that describes decision points and associated actions at those decision points.** The decision support matrix lists decision points, locations of decision points, criteria to be evaluated at decision points, actions that occur at decision points, and the units responsible to act on the decision points.

> ### Decision Making During Execution:
> ### Chamberlain at Little Round Top
>
> At 1630 on 2 July 1863, near Little Round Top, a rocky hill near Gettysburg, Pennsylvania, Colonel Joshua Chamberlain's 358 remaining soldiers of the 20th Maine Regiment were ordered into a defensive line. Minutes later, they came under a violent frontal assault by the 47th Alabama Regiment. While the 20th Maine was repulsing this assault, an officer rushed up to Chamberlain and informed him that another large enemy force was moving to attack their exposed left flank.
>
> Chamberlain immediately ordered a new defensive line at right angles to his existing line; he shifted the entire regiment to the left and back while maintaining continuous fire to the front and masking the movement of his left flank. Minutes later the 20th Maine was assaulted by the 15th Alabama Regiment. During that assault, the 20th Maine fired 20,000 rounds, suffering 30 percent dead and wounded. Chamberlain was bleeding and bruised. His foot bled from a shell fragment and his thigh was severely bruised after a musket ball had struck his scabbard. The 20th Maine miraculously withstood six charges before they ran out of ammunition.
>
> Chamberlain, fearing an overwhelming, decimating rebel attack, realized that by withdrawing he would be giving up key terrain as well as the battle. To the astonishment of his subordinates, he ordered a bayonet charge with the enemy, beginning another fierce charge from only 30 yards. The left half of his regiment began the charge, stunning the Confederates before them. As they came abreast of their own right half, Chamberlain raised high his saber and shouted, "Fix bayonets!" Running downhill, Chamberlain and his men had the clear advantage over the tired rebels. The Alabama men were shocked and fell back. A company of Chamberlain's men who had formed a screen line on the left flank began firing into the panic-stricken Confederates who, even though they outnumbered the 20th Maine 3 to 1, did not realize the strength of their numbers.
>
> Fearing the worst for his troops, Colonel William C. Oates, the commander of the Alabama regiments, ordered a breakout that turned into a rout and the capture of more than 400 of his men. Afterwards, Chamberlain was awarded the Medal of Honor. His actions serve as one of the finest examples of what a combat leader must be able to be and do to exercise effective mission command.

4-33. An *execution matrix* **is a visual representation of subordinate tasks in relationship to each other over time.** An operation can have multiple execution matrices. An execution matrix can cover the entire force for the duration of an operation; a specific portion of an operation (such as an air assault execution matrix); or for a specific warfighting function (such as a fire support execution matrix). Commanders and staffs use the execution matrix to control, synchronize, and adjust operations as required. An execution checklist is a distillation of the execution matrix that list key actions sequentially, units responsible for the action, and an associated code word to quickly provide shared understanding among the commander, staff, and subordinate units on initiation or completion of the action.

DIRECTING ACTION

4-34. To implement execution or adjustment decisions, commanders direct actions that apply combat power. Based on the commander's decision and guidance, the staff resynchronizes the operation to mass the maximum effects of combat power to seize, retain, and exploit the initiative. This involves synchronizing the operations in time, space, and purpose and issuing directives to subordinates. When modifying the plan, commander and staffs seek to—

- Make the fewest changes possible.
- Facilitate future operations.

4-35. Commanders make only those changes to the plan needed to correct variances. As much as possible, they keep the current plan the same to present subordinates with the fewest possible changes. Whenever possible, commanders ensure that changes do not preclude options for future operations. This is especially important for echelons above brigade.

RAPID DECISION-MAKING AND SYNCHRONIZATION PROCESS

4-36. The RDSP is a decision-making and planning technique that commanders and staffs commonly use during execution. While the MDMP seeks the optimal solution, the RDSP seeks a timely and effective solution within the commander's intent, mission, and concept of operations. Using the RDSP lets leaders avoid the time-consuming requirements of developing decision criteria and multiple COAs. Under the RDSP, leaders combine their experiences and intuition to quickly understand the situation, develop a viable option, and direct adjustments to the current order. When using this technique, the following considerations apply:

- Rapid is often more important than detailed analysis.
- Much of it may be mental rather than written.
- It should become a battle drill for the current operations integration cells, future operations cells, or both.

5-37. The RDSP is based on an existing order and the commander's priorities as expressed in the order. The most important of these are the commander's intent, concept of operations, and CCIRs. The RDSP includes five steps as shown in figure 4-3. Commanders perform the first two (as denoted in the oval) in any order, including concurrently. The last three (as denoted in the rectangle) are performed interactively until commanders identify a feasible, acceptable, and suitable COA.

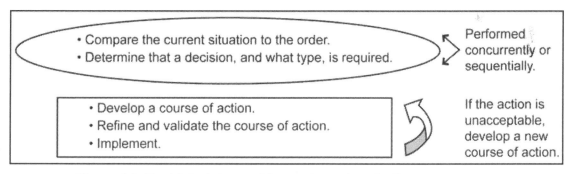

Figure 4-3. Rapid decision-making and synchronization process

4-38. The RDSP results in changes to the current order. Commanders often issue orders to subordinates verbally in situations requiring quick reactions. At battalion and higher levels, written FRAGORDs confirm verbal orders to ensure synchronization, integration, and notification of all parts of the force. If time permits, leaders verify that subordinates understand critical tasks. Methods for doing this include the confirmation brief and backbrief. (See FM 6-0 for a detailed discussion of the RDSP.)

This page intentionally left blank.

Chapter 5
Assessment

Estimation of the situation is, however, a continuous process, and changed conditions may, at any time, call for a new decision.

FM 100-5, *Operations* (1941)

This chapter defines assessment and describes its fundamentals. It describes an assessment process and concludes with guides for effective assessment.

FUNDAMENTALS OF ASSESSMENT

5-1. *Assessment* is the determination of the progress toward accomplishing a task, creating a condition, or achieving an objective (JP 3-0). Assessment is a continuous activity of the operations process that supports decision making by ascertaining progress of the operation for the purpose of developing and refining plans and for making operations more effective. Assessment results enhance the commander's decision making and help the commander and the staff to keep pace with constantly changing situations.

5-2. Assessment involves deliberately comparing intended outcomes with actual events to determine the overall effectiveness of force employment. More specifically, assessment helps the commander determine progress toward attaining the desired end state, achieving objectives, and performing tasks. Through professional military judgment, assessment helps answer the following questions:

- Where are we?
- What happened?
- Why do we think it happened?
- So what?
- What are the likely future opportunities and risks?
- What do we need to do?

5-3. Assessment precedes and guides the other activities of the operations process. During planning, assessment focuses on understanding an OE and building an assessment plan. During preparation, the focus of assessment switches to discerning changes in the situation and the force's readiness to execute operations. During execution, assessment involves deliberately comparing forecasted outcomes to actual events while using indicators to judge operational progress towards success. Assessment during execution helps commanders determine whether changes in the operation are necessary to take advantage of opportunities or to counter unexpected threats.

5-4. The situation and echelon dictate the focus and methods leaders use to assess. Assessment occurs at all echelons. Normally, commanders assess those specific operations or tasks that they were directed to accomplish. This properly focuses collection and assessment at each echelon, reduces redundancy, and enhances the efficiency of the overall assessment process.

5-5. For units with a staff, assessment becomes more formal at each higher echelon. Assessment resources (to include staff officer expertise and time available) proportionally increase from battalion to brigade, division, corps, and theater army. The analytic resources and level of expertise of staffs available at higher echelon headquarters include a dedicated core group of analysts. This group specializes in operations research and systems analysis, formal assessment plans, and various assessment products. Division, corps, and theater army headquarters, for example, have dedicated plans, future operations, and current operations integration cells. They have larger intelligence staffs and more staff officers trained in operations research and systems analysis. Assessment at brigade echelon and lower is usually less formal, often relying on direct observations and the judgment of commanders and their staffs.

5-6. For small units (those without a staff), assessment is mostly informal. Small-unit leaders focus on assessing their unit's readiness—personnel, equipment, supplies, and morale—and their unit's ability to perform assigned tasks. Leaders also determine whether the unit has attained task proficiency. If those tasks have not produced the desired results, leaders explore why they have not and consider what improvements could be made for unit operations. As they assess and learn, small units change their tactics, techniques, and procedures based on their experiences. In this way, even the lowest echelons in the Army follow the assessment process.

ASSESSMENT ACTIVITIES

5-7. The situation and type of operations affect the characteristics of assessment. During large-scale combat, assessment tends to be rapid, focused on the level of destruction of enemy units, terrain gained or lost, objectives secured, and the status of the friendly force to include sustainment. In other situations, such as counterinsurgency, assessment is less tangible. Assessing the level of security in an area or the level of the population's support for the government is challenging. Identifying what and how to assess requires significant effort from the commander and staff.

5-8. Whether conducting major combat operations or operations dominated by stability tasks, assessment consists of the major activities shown in figure 5-1. These activities include—

- Monitoring the current situation to collect relevant information.
- Evaluating progress toward attaining end state conditions, achieving objectives, and performing tasks.
- Recommending or directing action for improvement.

Figure 5-1. Activities of assessment

MONITORING

5-9. ***Monitoring* is continuous observation of those conditions relevant to the current operation.** Monitoring allows staffs to collect relevant information, specifically that information about the current situation described in the commander's intent and concept of operations. Commanders cannot judge progress nor make effective decisions without an accurate understanding of the current situation.

5-10. CCIRs and associated information requirements focus the staff's monitoring activities and prioritize the unit's collection efforts. Information requirements concerning the enemy, terrain and weather, and civil considerations are identified and assigned priorities through reconnaissance and surveillance. Operations officers use friendly reports to coordinate other assessment-related information requirements.

5-11. Staffs monitor and collect information from the common operational picture and friendly reports. This information includes operational and intelligence summaries from subordinate, higher, and adjacent headquarters and communications and reports from liaison teams. Staffs also identify information sources outside military channels and monitor their reports. These other channels might include products from civilian, host-nation, and other government agencies. Staffs apply information management and knowledge management to facilitate disseminating this information to the right people at the right time.

5-12. Staff sections record relevant information in running estimates. Staff sections maintain a continuous assessment of current operations as a basis to determine if operations are proceeding according to the commander's intent, mission, and concept of operations. In their running estimates, staff sections use this new information and these updated facts and assumptions as the basis for evaluation.

EVALUATING

> *While attrition formed an important element of American strategy in Vietnam, it hardly served as the guiding principle. Westmoreland believed that destroying enemy forces would help lead to larger political ends. Even before Westmoreland's tenure, MACV [Military Advisory Command Vietnam] had realized that quantitative reporting of enemy kills insufficiently measured progress in an unconventional environment. Body counts did not necessarily produce reliable qualitative assessments of the enemy's military and political strength.*
>
> Gregory A. Daddis

5-13. The staff analyzes relevant information collected through monitoring to evaluate the operation's progress. **Evaluating is using indicators to judge progress toward desired conditions and determining why the current degree of progress exists.** Evaluation is at the heart of the assessment process where most of the analysis occurs. Evaluation helps commanders determine what is working and what is not working, and it helps them gain insights into how to better accomplish the mission.

5-14. In the context of assessment, an *indicator* is a specific piece of information that infers the condition, state, or existence of something, and provides a reliable means to ascertain performance or effectiveness (JP 5-0). Indicators should be—

- Relevant—bear a direct relationship to a task, effect, object, or end state condition.
- Observable—collectable so that changes can be detected and measured or evaluated.
- Responsive—signify changes in the OE in time to enable effective decision making.
- Resourced—collection assets and staff resources are identified to observe and evaluate.

5-15. The two types of indicators commonly used in assessment include measures of performance (MOPs) and measures of effectiveness (MOEs). A *measure of performance* is an indicator used to measure a friendly action that is tied to measuring task accomplishment (JP 5-0). MOPs help answer questions such as "Was the action taken?" or "Were the tasks completed to standard?" A MOP confirms or denies that a task has been properly performed. MOPs are commonly found and tracked at all levels in execution matrixes. MOPs help to answer the question "Are we doing things right?"

5-16. At the most basic level, every Soldier assigned a task maintains a formal or informal checklist to track task completion. The status of those tasks and subtasks are MOPs. Similarly, operations consist of a series of collective tasks sequenced in time, space, and purpose to accomplish missions. Current operations integration cells use MOPs in execution matrixes, checklists, and running estimates to track completed tasks. Staffs use MOPs as a primary element of battle tracking with a focus on the friendly force. Evaluating task accomplishment using MOPs is relatively straightforward and often results in a "yes" or "no" answer.

5-17. A *measure of effectiveness* is an indicator used to measure a current system state, with change indicated by comparing multiple observations over time (JP 5-0). MOEs help measure changes in conditions, both positive and negative. MOEs help to answer the question "Are we doing the right things?" MOEs are commonly found and tracked in formal assessment plans.

5-18. Evaluation includes analysis of why progress is or is not being made. Commanders and staffs propose and consider possible causes. In particular, they address the question of whether or not changes in the situation can be attributed to friendly actions. Commanders consult subject matter experts, both internal and external to the staff, on whether their staffs have correctly identified the underlying causes for specific changes in the situation. These experts challenge key facts and assumptions identified in the planning process to determine if the facts and assumptions are still relevant or valid.

5-19. Evaluating also includes considering whether the desired conditions have changed, are no longer achievable, or are not achievable through the current operational approach. Staffs continually challenge the key assumptions made when framing the problem. When an assumption is invalidated, then reframing may be in order.

Chapter 5

RECOMMENDING OR DIRECTING ACTION

5-20. Monitoring and evaluating are critical activities; however, assessment is incomplete without recommending or directing action. Assessment may reveal problems, but unless it results in recommended adjustments, its use to the commander is limited. Ideally, recommendations highlight ways to improve the effectiveness of operations and plans by informing all decisions. (See paragraph 5-29 for a list of potential recommendations.)

5-21. Based on the evaluation of progress, the staff brainstorms possible improvements to the plan and makes preliminary judgments about the relative merit of those changes. Staff members identify those changes possessing sufficient merit and provide them as recommendations to the commander or make adjustments within their delegated authority. Recommendations to the commander range from continuing the operation as planned, to executing a branch, or to making unanticipated adjustments. Making adjustments includes assigning new tasks to subordinates, reprioritizing support, adjusting information collection assets, and significantly modifying the COA. Commanders integrate recommendations from the staff, subordinate commanders, and other partners with their personal assessments. Using those recommendations, they decide if and how to modify the operation to better accomplish the mission.

5-22. Assessment helps identify threats, suggests improvements to effectiveness, and reveals opportunities. The staff presents the results and conclusions of its assessments and recommendations to the commander as an operation develops. Just as the staff devotes time to analysis and evaluation, so too must it make timely, complete, and actionable recommendations. The chief of staff or executive officer ensures the staff completes its analyses and recommendations in time to affect the operation and for information to reach the commander when needed.

Measures of Effectiveness: OPERATION SUPPORT HOPE

Mission statements can serve as a primary source from which to develop measures of effectiveness. The first and most urgent task facing planners for OPERATION SUPPORT HOPE in Rwanda, July 1994, was the directive to "stop the dying" from the U.S. Commander in Chief, Europe. Initial action focused on the massive refugee deaths from cholera around Goma, Zaire. The joint task force commander decided to measure effectiveness by whether refugee camp death rates dropped to the level the United Nations determined was consistent with "normal" camp operations. A related mission requirement was to open Kigali airfield for 24-hour operations. Success for this requirement was measured by statistical data that showed the surge in airfield use and cargo throughput after American forces arrived. Both measures of effectiveness derived from the mission statement were used throughout the operation to communicate progress to all participants.

ASSESSMENT PROCESS

5-23. There is no single way to conduct assessment. Every situation has its own distinctive challenges, making every assessment unique. The following steps can help guide the development of an effective assessment plan and assessment activities during preparation and execution:
- Step 1 – Develop the assessment approach (planning).
- Step 2 – Develop the assessment plan (planning).
- Step 3 – Collect information and intelligence (preparation and execution).
- Step 4 – Analyze information and intelligence (preparation and execution).
- Step 5 – Communicate feedback and recommendations (preparation and execution).
- Step 6 – Adapt plans or operations (planning and execution).

(See ATP 5-0.3 for a detailed discussion of each step of the assessment process.)

STEP 1 – DEVELOP THE ASSESSMENT APPROACH

5-24. Assessment begins in planning as the commander identifies the operation's end state, operational approach, and associated objectives and tasks. Concurrently, the staff begins to develop an assessment approach by identifying specific information needed to monitor and analyze conditions associated with attaining the operation's end state, achieving objectives, and accomplishing tasks. In doing so, the staff tries to answer the following questions:
- How will we know we are creating the desired conditions?
- What information do we need?
- Who is best postured to provide that information?

5-25. If a higher headquarters assessment plan exists, the staff aligns applicable elements of that assessment plan to the plan they are developing. The assessment approach becomes the framework for the assessment plan and will continue to mature through plan development. The assessment approach should identify the information and intelligence needed to assess progress and inform decision making.

STEP 2 – DEVELOP THE ASSESSMENT PLAN

5-26. This step overlaps Step 1. It focuses on developing a plan to monitor and collect necessary information and intelligence to inform decision making throughout execution. The assessment plan should link end state conditions, objectives, and tasks to observable key indicators. This plan also should include specific staff responsibilities to monitor, collect, and analyze information as well as develop recommendations and assessment products as required.

STEP 3 – COLLECT INFORMATION

5-27. Staffs collect relevant information throughout planning and execution. They refine and adapt information collection requirements as the operations progresses. Staffs and subordinate commands provide information during execution through applicable battle rhythm events and reports. Intelligence staffs continually provide updates about the situation to include information about the enemy, terrain, and civil considerations.

STEP 4 – ANALYZE INFORMATION AND INTELLIGENCE

5-28. Analysis seeks to identify positive or negative movement toward achieving objectives or attaining end state conditions. Accurate analysis seeks to identify trends and changes that significantly impact the operation. Based on this analysis, the staff estimates the effects of force employment and resource allocation; determines whether forces have achieved their objectives; or realizes that a decision point has emerged.

5-29. Recommendations generated by staff analyses regarding achievement of the objective or attainment of the desired end state conditions, force employment, resource allocation, validity of planning assumptions, and decision points should enable the staff to develop recommendations for consideration. Recommendations can include the following:
- Update, change, add, or remove critical assumptions.
- Transition between phases.
- Execute branches or sequels.
- Change resource allocation.
- Adjust objectives or end state conditions.
- Change or add tasks to subordinate units.
- Adjust priorities.
- Change priorities of effort.
- Change command relationships.
- Change task organizations.
- Adjust decision points.
- Refine or adapt the assessment plan.

STEP 5 – COMMUNICATE FEEDBACK AND RECOMMENDATIONS

5-30. Assessment products contain recommendations for the commander based upon the commander's guidance. Regardless of quality and effort, the assessment process is limited if the communication of its results is deficient or inconsistent with the commander's personal style of assimilating information and making decisions. Additionally, there may be a requirement to provide input to higher headquarters assessments in which the requirements and feedback could be within a different construct.

STEP 6 – ADAPT PLANS OR OPERATIONS

5-31. Commanders direct changes or provide additional guidance that dictate updates or modifications to operations to drive progress of operations to objectives and end state conditions. Staffs capture the commander's decisions and guidance to ensure forces take necessary actions. As the operation evolves, the assessment plan will evolve as well.

GUIDES TO EFFECTIVE ASSESSMENT

5-32. Throughout the conduct of operations, commanders integrate their own assessments with those of the staff, subordinate commanders, and other unified action partners in the AO. The following guides aid in effective assessment:
- Commander involvement.
- Integration.
- Incorporation of the logic of the plan.
- Caution when establishing cause and effect.

COMMANDER INVOLVEMENT

5-33. The commander's involvement in operation assessment is essential. The assessment plan should focus on the information and intelligence that directly support the commander's decision making. Commanders establish priorities for assessment in their planning guidance and CCIRs. By prioritizing the effort, commanders guide the staff's analysis efforts. Committing valuable time and energy to developing excessive and time-consuming assessment schemes squanders resources better devoted to other operations process activities. Commanders reject the tendency to measure something just because it is measurable. Effective commanders avoid burdening subordinates and staffs with overly detailed assessments and collection tasks. Generally, the echelon at which a specific operation, task, or action is conducted should be the echelon at which it is assessed.

> **Commander's Assessment: Are We Ready to Close on Baghdad**
>
> Just eight days after crossing the border from Kuwait into Iraq, coalition forces neared the enemy's outer defensive positions of Baghdad. V Corps had fought its way north from the west side of the Euphrates River as the 1st Marine Expeditionary Force (MEF) attacked north on the east side. Both forces had weathered a sand storm and logisticians diligently moved supplies, fuel, and ammunition forward. Fighting continued along the V Corps line of operations from As Samawah north to An Najaf. Intense fighting was ongoing along the I MEF's approach from Basra north to An Nasiriyah.
>
> On 28 March 2003, General David D. McKiernan, commander of the coalition force land component command, went forward to meet with his commanders, Generals James T. Conway (I MEF) and William S. Wallace (V Corps) in Jalibah. He wanted to hear directly from his subordinates concerning their "stance" for the transition from the march up-country to closing on Baghdad.
>
> The meeting began with McKiernan providing his assessment of enemy forces and results of war gaming from his staff. He then asked some key questions to his commanders, including their satisfaction with the level of risk along the lines of communications back to Kuwait. Both Wallace and Conway had some concerns they believed they needed to address prior to crossing the "red line" that referred to entering the inner defensive cordon outside of Baghdad. Wallace briefed his plan for a series of attacks designed to set the conditions for the assault to isolate Baghdad. McKiernan asked what he needed to set that stance. Wallace responded by saying he needed to position the corps by 31 March to launch his attacks on 1 April.
>
> Conway noted that the I MEF was undertaking "a systematic reduction of the bad guys in An Nasiriya" and he wanted 1 United Kingdom Armored Division to execute some "pinpoint armor strikes" in Basra. Conway also noted that Colonel Joe Dowdy (1st Regimental Combat Team commander) was in "a 270-degree fight."
>
> After hearing his commanders, McKiernan decided to "take time to clean up and make sure we have the right stance in our battlespace before we commit into the Baghdad fight, because once we commit to the Baghdad fight, we can't stop." Supported by the assessment of his commanders, McKiernan ordered a transition to set conditions to isolate Baghdad.

INTEGRATION

5-34. Assessment requires integration. Assessing progress is the responsibility of all staff sections and not the purview of any one staff section or command post cell. Each staff section assesses the operation from its specific area of expertise. However, these staff sections must coordinate and integrate their individual assessments and associated recommendations across the warfighting functions to produce comprehensive assessments for the commander, particularly in protracted operations. They do this in the assessment working group.

5-35. Most assessment working groups are at higher echelons (division and above) and are more likely to be required in protracted operations. Normally, the frequency of meetings is part of a unit's battle rhythm. The staff, however, does not wait for a scheduled working group to inform the commander on issues that require immediate attention. Nor does the staff wait to take action in those areas within its delegated authority.

5-36. The assessment working group is cross-functional by design and includes membership from across the staff, liaison personnel, and other unified action partners outside the headquarters. Commanders direct the chief of staff, executive officer, or a staff section leader to run the assessment working group. Typically, the operations officer, plans officer, or senior operations research and systems analysis staff section serves as the staff lead for the assessment working group.

5-37. Developing an assessment plan occurs concurrently within the steps of the MDMP. The resulting assessment plan should support the command's battle rhythm. The frequency with which the assessment working group meets depends on the situation. Additionally, the assessment working group may present its findings and recommendations to the commander for decision. Subordinate commanders may participate and provide their assessments of operations and recommendations along with the staff. Commanders combine these assessments with their personal assessment, consider recommendations, and then direct changes to improve performance and better accomplish the mission.

INCORPORATION OF THE LOGIC OF THE PLAN

5-38. Effective assessment relies on an accurate understanding of the logic (reasoning) used to build the plan. Each plan is built on assumptions and an operational approach. The reasons or logic as to why the commander believes the plan will produce the desired results become important considerations when staffs determine how to assess operations. Recording, understanding, and making this logic explicit helps the staffs recommend the appropriate MOEs and MOPs for assessing the operation.

CAUTION WHEN ESTABLISHING CAUSE AND EFFECT

5-39. Although establishing cause and effect is sometimes difficult, it is crucial to effective assessment. Sometimes, establishing causality between actions and their effects can be relatively straightforward, such as in observing a bomb destroy a bridge. In other instances, especially regarding changes in human behavior, attitudes, and perception, establishing links between cause and effect proves difficult. Commanders and staffs must guard against drawing erroneous conclusions in these instances.

Source Notes

This division lists sources by page number. Where material appears in a paragraph, it lists both the page number followed by the paragraph number.

1-1 "The best is…": General George S. Patton, *War as I Knew It* (Boston, MA: Houghton Mifflin Company, 1947), 354.

1-2 "Everything in war…": Carl von Clausewitz, *On War* (Boston, MA: Princeton University Press, 1976), 119.

1-2 "Thus any study…": B. H. Liddell Hart, *Strategy* (New York, NY: Signet Printing, 1974), 351.

1-3 "Diverse are the…": Field Marshal Helmuth von Moltke cited in Major General Werner Widder, "Auftragstaktik and Innere Fuhrung: Trademarks of German Leadership," *Military Review* (September-October 2002 English Edition): 4.

1-5 **Agility: Rapidly Turning the Third Army to Bastogne**, Combined Arms Doctrine Directorate staff, unpublished text, 2018. Based on George S. Patton, Jr., *War as I Knew It* (Boston, MA: Houghton Mifflin Company, 1947), 189–191.

1-6 "It is a mistake…": Bernard L. Montgomery, *The Memoirs of Field-Marshal the Viscount Montgomery of Alamein, K.G.* (Cleveland, OH: The World Publishing Company, 1958), 76.

1-9 "I suppose dozens…": Field-Marshall Viscount William Slim, *Defeat into Victory: Battling Japan in Burma and India, 1942–1945* (New York, NY: Copper Square Press, 2000), 210.

1-12 "Example whether it…": George Washington. "From George Washington to Major General Stirling, 5 March 1780," Founders Online, National Archives, accessed April 11, 2019, https://founders.archives.gov/documents/Washington/03-24-02-0525. [Original source: *The Papers of George Washington, Revolutionary War Series*, vol. 24, 1 January–9 March 1780, ed. Benjamin L. Huggins. Charlottesville, VA: University of Virginia Press, 2016, pp. 630–632.]

1-13 "If you know…": Sun Tzu, *The Art of War*, trans. Lionel Giles (1910), 24–25.

1-14 **Collaboration: Meade's Council of War**. Adapted from Steve Dundas' blog, "A Council of War: Meade and His Generals Decide to Stay and Fight at Gettysburg July 2nd 1863." Blog, Padre Steve's World (https://padresteve.com/2014/04/25/a-council-of-war-meade-and-his-generals-decide-to-stay-and-fight-at-gettysburg-july-2nd-1863/). Accessed 02 April 2018. Dundas cites quotations from Halleck, Meade, and Butterfield found in Stephen W. Sears' *Gettysburg* (New York, NY: Houghton Mifflin Company, 2003), 341–343.

2-1 "To be practical…": B. H. Liddell Hart, *Strategy* (New York, NY: Signet Printing, 1974), 343–344.

2-2 "Logistics comprises the…": Antoine Henri de Jomini, *Art of War*, translated by G.H. Mendell and W.P. Craighill (Philadelphia, PA: J.B. Lippincott & Co., 1862; The Internet Archive, 22 February 2009), 69. https://archive.org/details/artwar00mendgoog.

2-3 "I tell this story…": Dwight D. Eisenhower, "Remarks at the National Defense Executive Reserve Conference," November 14, 1957. Online by Gerhard Peters and John T. Woolley, *The American Presidency Project*. http://www.presidency.ucsb.edu/ws/?pid=10951.

2-6 "In general, campaign…": Frederick the Great as cited in Owen Connolly, *On War and Leadership* (Princeton, NJ: Princeton University Press, 2002), 15.

2-7 "War plans cover…": Carl von Clausewitz, *On War* (Boston, MA: Princeton University Press, 1976), 579.

Source Notes

2-10 "Nothing succeeds in...": Napoleon Bonaparte, *Dictionary of Military and Naval Quotations*, compiled by Robert Debs Heinl, Jr. (Annapolis, MD: United States Naval Institute, 1967), 239.

2-12 "For Alexander, Gustavus...": Carl von Clausewitz, *On War* (Boston, MA: Princeton University Press, 1976), 596.

2-13 "If the art...": Antoine Henri de Jomini, *Art of War* (St. Paul, MN: Greenhill Books, 2006), 114.

2-15 "It is my experience...": Field Marshal Erwin Rommel cited in Owen Connolly, *On War and Leadership* (Princeton, NJ: Princeton University Press, 2002), 106.

2-18 "An order should...": *Field Service Regulations: United States Army* (obsolete) (Washington, DC: Government Printing Office, 1905), 29.

2-20 "Now the general...": Sun Tzu, *The Art of War*, trans. Lionel Giles (1910), 7.

2-22 **Tenets in Action: OPERATION JUST CAUSE.** FM 100-5, *Operations* (obsolete) (Washington, DC: Government Printing Office, 1993), 6-3–6-4.

2-22 "It is my...": General George S. Patton, *War as I Knew It* (Boston, MA: Houghton Mifflin Company, 1947), 357.

2-23 "You can ask...": Napoleon Bonaparte, *Dictionary of Military and Naval Quotations*, compiled by Robert Debs Heinl, Jr. (Annapolis, MD: United States Naval Institute, 1967), 325.

2-24 "The process of...": FM 101-5, *Staff Officers' Field Manual: The Staff and Combat Orders* (obsolete) (Washington, DC: Government Printing Office, 1940), 37.

2-25 "Since all information...": Carl von Clausewitz, *On War*, translated and edited by M. Howard and P. Paret (Boston, MA: Princeton University Press, 1976), 102.

2-25 "In war, leaders...": *Infantry In Battle* (obsolete) (Washington, DC: The Infantry Journal Incorporated, 1939), 161.

3-1 "The stroke of...": Marshal Ferdinand Foch, *Dictionary of Military and Naval Quotations*, compiled by Robert Debs Heinl, Jr. (Annapolis, MD: United States Naval Institute, 1967), 239.

3-1 "Before the battle...": Bernard Law Montgomery cited in Owen Connolly, *On War and Leadership* (Princeton, NJ: Princeton University Press, 2002), 153.

3-2 "Nine-tenths of tactics...": T. E. Lawrence, "The Evolution of a Revolt," *The Army Quarterly and Defence Journal*, Volume 1 (London: William Clowes & Sons, Ltd., October 1920 and January 1921), 60.

3-3 **Prepare: Rangers Train for Seizing Pointe du Hoc.** JoAnna M. McDonald, *The Liberation of Pointe du Hoc: the 2nd U.S. Rangers at Normandy* (Redondo Beach, CA: Rank and File Publications, 2000).

3-4 "If I always...": Napoleon Bonaparte, *Dictionary of Military and Naval Quotations*, compiled by Robert Debs Heinl, Jr. (Annapolis, MD: United States Naval Institute, 1967), 239.

3-6 "In no other...": General Douglas MacArthur, *Dictionary of Military and Naval Quotations*, compiled by Robert Debs Heinl, Jr. (Annapolis, MD: United States Naval Institute, 1967), 329.

3-7 "Sand-Table Exercises...": General George S. Patton, *War as I Knew It* (Boston, MA: Houghton Mifflin Company, 1947), 354.

3-9 **Large-Unit Preparation: Third Army Readies for OPERATION IRAQI FREEDOM.** Adapted from Gregory Fontenot, *On Point: United States Army in Operation Iraqi Freedom* (Fort Leavenworth, KS: Combat Studies Institute Press, 2004), 29–84.

4-1 "No plan of...": Helmuth von Moltke, *Moltke's Military Works*, Vol. 4, War Lessons, Part I, "Operative Preparations for Battle," translated by Harry Bell (Fort Leavenworth, KS: Army Service Schools, 1916), 66.

4-1	4-2 "…one makes plans…": George S. Patton Jr., *The Patton Papers, vol. 2, 1940–1945*, ed. Martin Blumenson (Boston, MA: Houghton Mifflin Co., 1974), 648.
4-1	"I am heartily…": Ulysses S. Grant quoted in Horace Porter, *Campaigning with Grant* (New York, NY: The Century Co., 1907), 70.
4-2	"I have found…": Erwin Rommel, *The Rommel Papers*, edited by B. H. Liddell-Hart (New York, NY: Harcourt, Brace, and Company, 1953), 7.
4-3	"To be at the…": General William Tecumseh Sherman, *Memoirs of General W. T. Sherman* (New York, NY: D. Appleton & Company, 1891; The Project Gutenberg, 10 June 2004 [EBook #5853]), 407. https://ia802605.us.archive.org/23/items/thememoirsofgene05853gut/5853.txt.
4-4	"The commander's mission…": FM 100-5, *Field Service Regulations: Operations* (obsolete) (Washington, DC: Government Printing Office, 1941), 24.
4-5	"In war obscurity…": *Infantry In Battle* (obsolete) (Washington, DC: The Infantry Journal Incorporated, 1939), 16.
4-7	**Decision Making During Execution: Chamberlain at Little Round Top**. Vignette adapted from Joshua Lawrence Chamberlain, "Through Blood and Fire at Gettysburg: General Joshua Chamberlain and the 20th Maine," *Hearst's Magazine* Vol. 23 (New York, NY: Charles Schweinier Press, January 1913), 894–909. Available at http://www.joshualawrencechamberlain.com/bloodandfire.php. Also adapted from H. S. Melcher, "The 20th Maine at Little Round Top," *Battles and Leaders of the Civil War* Vol. 3, Condensed from the "Lincoln County News," Waldoboro, Maine, March 13th, 1885. Available at http://www.joshualawrencechamberlain.com/20me7.php.
5-1	"Estimation of the…": FM 100-5, *Field Service Regulations: Operations* (obsolete) (Washington, DC: Government Printing Office, 1941), 26.
5-3	"While attrition formed…": Gregory A. Daddis, *No Sure Victory: Measuring U.S. Army Effectiveness and Progress in the Vietnam War* (New York, NY: Oxford University Press, 2011), 7.
5-5	**Measures of Effectiveness: OPERATION SUPPORT HOPE**. Adapted from John E. Lange "Civilian-Military Cooperation and Humanitarian Assistance: Lessons from Rwanda," *Parameters* (Summer 1998): 106–122.
5-6	**Commander's Assessment: Are We Ready to Close on Baghdad**. Adapted from Gregory Fontenot, *On Point: United States Army in Operation Iraqi Freedom* (Fort Leavenworth, KS: Combat Studies Institute Press, 2004), 245.

This page intentionally left blank.

Glossary

The glossary lists acronyms and terms with Army or joint definitions. Where Army and joint definitions differ, (Army) precedes the definition. Terms for which ADP 5-0 is the proponent are marked with an asterisk (*). The proponent publication for other terms is listed in parentheses after the definition.

SECTION I – ACRONYMS AND ABBREVIATIONS

ADM	Army design methodology
ADP	Army doctrine publication
ADRP	Army doctrine reference publication
AO	area of operations
ATP	Army techniques publication
CCIR	commander's critical information requirement
CCP	combatant command campaign plan
CJCSM	Chairman of the Joint Chiefs of Staff manual
COA	course of action
DA	Department of the Army
FM	field manual
FRAGORD	fragmentary order
GCC	geographic combatant commander
IPB	intelligence preparation of the battlefield
JP	joint publication
LNO	liaison officer
MDMP	military decision-making process
MOE	measure of effectiveness
MOP	measure of performance
OE	operational environment
OPLAN	operation plan
OPORD	operation order
RDSP	rapid decision-making and synchronization process
SOP	standard operating procedure
TLP	troop leading procedures

Glossary

| U.S. | United States |
| WARNORD | warning order |

SECTION II – TERMS

*Army design methodology
  A methodology for applying critical and creative thinking to understand, visualize, and describe problems and approaches to solving them.

assessment
  The determination of the progress toward accomplishing a task, creating a condition, or achieving an objective. (JP 3-0)

battle rhythm
  A deliberate, daily schedule of command, staff, and unit activities intended to maximize use of time and synchronize staff actions. (JP 3-33)

branch
  The contingency options built into the base plan used for changing the mission, orientation, or direction of movement of a force to aid success of the operation based on anticipated events, opportunities, or disruptions caused by enemy actions and reactions. (JP 5-0)

campaign plan
  A joint operation plan for a series of related major operations aimed at achieving strategic or operational objectives within a given time and space. (JP 5-0)

center of gravity
  The source of power that provides moral or physical strength, freedom of action, or will to act. (JP 5-0)

*collaborative planning
  Two or more echelons planning together in real time, sharing information, perceptions, and ideas to develop their respective plans simultaneously.

combat power
  (Army) The total means of destructive, constructive, and information capabilities that a military unit or formation can apply at a given time. (ADP 3-0)

command and control
  The exercise of authority and direction by a properly designated commander over assigned and attached forces in the accomplishment of the mission. (JP 1)

commander's critical information requirement
  An information requirement identified by the commander as being critical to facilitating timely decision making. (JP 3-0)

commander's intent
  A clear and concise expression of the purpose of the operation and the desired military end state that supports mission command, provides focus to the staff, and helps subordinate and supporting commanders act to achieve the commander's desired results without further orders, even when the operation does not unfold as planned. (JP 3-0)

commander's visualization
  The mental process of developing situational understanding, determining a desired end state, and envisioning an operational approach by which the force will achieve that end state. (ADP 6-0)

*confirmation brief
  A brief subordinate leaders give to the higher commander immediately after the operation order is given to confirm understanding.

*concept of operations

(Army) A statement that directs the manner in which subordinate units cooperate to accomplish the mission and establishes the sequence of actions the force will use to achieve the end state.

control measure

A means of regulating forces or warfighting functions. (ADP 6-0)

coordinated fire line

A line beyond which conventional surface-to-surface direct fire and indirect fire support means may fire at any time within the boundaries of the establishing headquarters without additional coordination but does not eliminate the responsibility to coordinate the airspace required to conduct the mission. (JP 3-09)

culminating point

The point at which a force no longer has the capability to continue its form of operations, offense or defense. (JP 5-0)

decision point

A point in space and time when the commander or staff anticipates making a key decision concerning a specific course of action. (JP 5-0)

*decision support matrix

A written record of a war-gamed course of action that describes decision points and associated actions at those decision points.

decision support template

A combined intelligence and operations graphic based on the results of wargaming that depicts decision points, timelines associated with movement of forces and the flow of the operation, and other key items of information required to execute a specific friendly course of action. (JP 2-01.3)

decisive point

A geographic place, specific key event, critical factor, or function that, when acted upon, allows commanders to gain a marked advantage over an enemy or contribute materially to achieving success. (JP 5-0)

depth

The extension of operations in time, space, or purpose to achieve definitive results. (ADP 3-0)

end state

The set of required conditions that defines achievement of the commander's objectives. (JP 3-0)

essential element of friendly information

A critical aspect of a friendly operation that, if known by a threat would subsequently compromise, lead to failure, or limit success of the operation and therefore should be protected from enemy detection. (ADP 6-0)

*evaluating

Using indicators to judge progress toward desired conditions and determining why the current degree of progress exists.

*execution

The act of putting a plan into action by applying combat power to accomplish the mission and adjusting operations based on changes in the situation.

*execution matrix

A visual representation of subordinate tasks in relationship to each other over time.

flexibility

The employment of a versatile mix of capabilities, formations, and equipment for conducting operations. (ADP 3-0)

friendly force information requirement

Information the commander and staff need to understand the status of friendly force and supporting capabilities. (JP 3-0)

graphic control measure

A symbol used on maps and displays to regulate forces and warfighting functions. (ADP 6-0)

indicator

In the context of assessment, a specific piece of information that infers the condition, state, or existence of something, and provides a reliable means to ascertain performance or effectiveness. (JP 5-0)

information collection

An activity that synchronizes and integrates the planning and employment of sensors and assets as well as the processing, exploitation, and dissemination systems in direct support of current and future operations. (FM 3-55)

intelligence preparation of the battlefield

The systematic process of analyzing the mission variables of enemy, terrain, weather, and civil considerations in an area of interest to determine their effect on operations. (ATP 2-01.3)

key tasks

Those activities the force must perform as a whole to achieve the desired end state. (ADP 6-0)

knowledge management

The process of enabling knowledge flow to enhance shared understanding, learning, and decision making. (ADP 6-0)

leadership

The activity of influencing people by providing purpose, direction, and motivation to accomplish the mission and improve the organization. (ADP 6-22)

levels of warfare

A framework for defining and clarifying the relationship among national objectives, the operational approach, and tactical tasks. (ADP 1-01)

line of effort

(Army) A line that links multiple tasks using the logic of purpose rather than geographical reference to focus efforts toward establishing a desired end state. (ADP 3-0)

line of operations

A line that defines the directional orientation of a force in time and space in relation to the enemy and links the force with its base of operations and objectives. (ADP 3-0)

main effort

A designated subordinate unit whose mission at a given point in time is most critical to overall mission success. (ADP 3-0)

measure of effectiveness

An indicator used to measure a current system state, with change indicated by comparing multiple observations over time. (JP 5-0)

measure of performance

A indicator used to measure a friendly action that is tied to measuring task accomplishment. (JP 5-0)

*military decision-making process

An iterative planning methodology to understand the situation and mission, develop a course of action, and produce an operation plan or order.

mission

    The task, together with the purpose, that clearly indicates the action to be taken and the reason therefore. (JP 3-0)

mission command

    (Army) The Army's approach to command and control that empowers subordinate decision making and decentralized execution appropriate to the situation. (ADP 6-0)

mission orders

    Directives that emphasize to subordinates the results to be attained, not how they are to achieve them. (ADP 6-0)

*monitoring

    Continuous observation of those conditions relevant to the current operation.

multinational operations

    A collective term to describe military actions conducted by forces of two or more nations, usually undertaken within the structure of a coalition or alliance. (JP 3-16).

*nested concepts

    A planning technique to achieve unity of purpose whereby each succeeding echelon's concept of operations is aligned by purpose with the higher echelons' concept of operations.

objective

    The clearly defined, decisive, and attainable goal toward which an operation is directed. (JP 5-0)

operational approach

    A broad description of the mission, operational concepts, tasks, and actions required to accomplish the mission. (JP 5-0)

operational art

    The cognitive approach by commanders and staffs—supported by their skill, knowledge, experience, creativity, and judgment—to develop strategies, campaigns, and operations to organize and employ military forces by integrating ends, ways, and means. (JP 3-0)

operational environment

    A composite of the conditions, circumstances, and influences that affect the employment of capabilities and bear on the decisions of the commander. (JP 3-0)

operational level of warfare

    The level of warfare at which campaigns and major operations are planned, conducted, and sustained to achieve strategic objectives within theaters or other operational areas. (JP 3-0)

operational reach

    The distance and duration across which a force can successfully employ military capabilities. (JP 3-0)

*operations process

    The major command and control activities performed during operations: planning, preparing, executing, and continuously assessing the operation.

*parallel planning

    Two or more echelons planning for the same operations nearly simultaneously facilitated by the use of warning orders by the higher headquarters.

phase

    (Army) A planning and execution tool used to divide an operation in duration or activity. (ADP 3-0)

*planning

    The art and science of understanding a situation, envisioning a desired future, and determining effective ways to bring that future about.

*planning horizon

A point in time commanders use to focus the organization's planning efforts to shape future events.

*preparation

Those activities performed by units and Soldiers to improve their ability to execute an operation.

priority intelligence requirement

An intelligence requirement that the commander and staff need to understand the threat and other aspects of the operational environment. (JP 2-01)

*priority of support

A priority set by the commander to ensure a subordinate unit has support in accordance with its relative importance to accomplish the mission.

*rehearsal

A session in which the commander and staff or unit practices expected actions to improve performance during execution.

risk management

The process to identify, assess, and control risks and make decisions that balance risk cost with mission benefits. (JP 3-0)

route

The prescribed course to be traveled from a specific point of origin to a specific destination. (FM 3-90-1)

*running estimate

The continuous assessment of the current situation used to determine if the current operation is proceeding according to the commander's intent and if planned future operations are supportable.

sequel

The subsequent operation or phase based on the possible outcomes of the current operation or phase. (JP 5-0)

simultaneity

The execution of related and mutually supporting tasks at the same time across multiple locations and domains. (ADP 3-0)

situational understanding

The product of applying analysis and judgment to relevant information to determine the relationships among the operational and mission variables. (ADP 6-0)

strategic level of warfare

The level of warfare at which a nation, often as a member of a group of nations, determines national or multinational (alliance or coalition) strategic security objectives and guidance, then develops and uses national resources to achieve those objectives. (JP 3-0)

synchronization

The arrangement of military actions in time, space, and purpose to produce maximum relative combat power at a decisive place and time. (JP 2-0)

tactical level of warfare

The level of warfare at which battles and engagements are planned and executed to achieve military objectives assigned to tactical units or task forces. (JP 3-0)

targeting

The process of selecting and prioritizing targets and matching the appropriate response to them, considering operational requirements and capabilities. (JP 3-0)

*task organization

(Army) A temporary grouping of forces designed to accomplish a particular mission.

task-organizing

The act of designing a force, support staff, or sustainment package of specific size and composition to meet a unique task or mission. (ADP 3-0)

tempo

The relative speed and rhythm of military operations over time with respect to the enemy. (ADP 3-0)

tenets of operations

Desirable attributes that should be built into all plans and operations and are directly related to the Army's operational concept. (ADP 1-01)

*troop leading procedures

A dynamic process used by small-unit leaders to analyze a mission, develop a plan, and prepare for an operation.

unified land operations

Simultaneous execution of offense, defense, stability, and defense support of civil authorities across multiple domains to shape operational environments, prevent conflict, prevail in large-scale ground combat, and consolidate gains as part of unified action. (ADP 3-0)

This page intentionally left blank.

References

All URLs accessed on 17 June 2019.

REQUIRED PUBLICATIONS

Readers require these publications for fundamental concepts, terms, and definitions.

DOD Dictionary of Military and Associated Terms. June 2019.

ADP 1-02. *Terms and Military Symbols*. 14 August 2018.

ADP 3-0. *Operations*. 31 July 2019.

RELATED PUBLICATIONS

These publications are referenced in this publication.

JOINT PUBLICATIONS

Joint publications and Chairman of the Joint Chiefs of Staff directives are available at https://www.jcs.mil/Doctrine/.

CJCSM 3130.03A. *Planning and Execution Planning Formats and Guidance*.

JP 1. *Doctrine for the Armed Forces of the United States*. 25 March 2013.

JP 2-0. *Joint Intelligence*. 22 October 2013.

JP 2-01. *Joint and National Intelligence Support to Military Operations*. 05 July 2017.

JP 2-01.3. *Joint Intelligence Preparation of the Operational Environment*. 21 May 2014.

JP 3-0. *Joint Operations*. 17 January 2017.

JP 3-09. *Joint Fire Support*. 10 April 2019.

JP 3-16. *Multinational Operations*. 01 March 2019.

JP 3-31. *Joint Land Operations*. 24 February 2014.

JP 3-33. *Joint Task Force Headquarters*. 31 January 2018.

JP 3-34. *Joint Engineer Operations*. 06 January 2016.

JP 5-0. *Joint Planning*. 16 June 2017.

ARMY PUBLICATIONS

Army doctrinal publications are available at https://armypubs.army.mil/.

ADP 1-01. *Doctrine Primer*. 31 July 2019.

ADP 2-0. *Intelligence*. 31 July 2019.

ADP 3-37. *Protection*. 31 July 2019.

ADP 3-90. *Offense and Defense*. 31 July 2019.

ADP 6-0. *Mission Command: Command and Control of Army Forces*. 31 July 2019

ADP 6-22. *Army Leadership*. 31 July 2019.

ATP 2-01.3. *Intelligence Preparation of the Battlefield*. 01 March 2019.

ATP 3-37.10/MCRP 3-40D.13. *Base Camps*. 27 January 2017.

ATP 3-60. *Targeting*. 07 May 2015.

ATP 5-0.1. *Army Design Methodology*. 01 July 2015.

References

ATP 5-0.3/MCRP 5-1C/NTTP 5-01.3/AFTTP 3-2.87. *Multi-Service Tactics, Techniques, and Procedures for Operation Assessment*. 18 August 2015.

ATP 5-19. *Risk Management*. 14 April 2014.

ATP 6-0.5. *Command Post Organization and Operations*. 01 March 2017.

ATP 6-01.1. *Techniques for Effective Knowledge Management*. 06 March 2015.

FM 3-0. *Operations*. 06 October 2017.

FM 3-16. *The Army in Multinational Operations*. 08 April 2014.

FM 3-34. *Engineer Operations*. 02 April 2014.

FM 3-55. *Information Collection*. 03 May 2013.

FM 3-90-1. *Offense and Defense, Volume 1*. 22 March 2013.

FM 6-0. *Commander and Staff Organization and Operations*. 05 May 2014.

FM 27-10. *The Law of Land Warfare*. 18 July 1956.

OBSOLETE PUBLICATIONS

This section contains references to obsolete historical doctrine. The Archival and Special Collections in the Combined Arms Research Library (CARL), Fort Leavenworth, Kansas contains copies. These publications are obsolete doctrine publications referenced for citations only.

Field Service Regulations: United States Army. Washington, DC: Government Printing Office, 1905.

FM 101-5. *Staff Officers Field Manual*. Washington, DC: Government Printing Office, 1940.

FM 100-5. *Field Service Regulations: Operations*. Washington, DC: Government Printing Office, 1941.

FM 100-5. *Operations*. Washington, DC: Government Printing Office, 1993.

OTHER PUBLICATIONS

This section contains other references. All websites accessed 28 June 2018.

Allied Tactical Publication 3.2.2. *Command and Control of Allied Forces*. 15 December 2016.

Blumenson, Martin ed. *The Patton Papers*. Vol. 2. 1940–1945. Boston, MA: Houghton Mifflin Co., 1974.

Chamberlain, Joshua Lawrence. "Through Blood and Fire at Gettysburg: General Joshua Chamberlain and the 20th Maine." *Hearst's Magazine*. Vol. 23. January 1913, 894–909. Available at http://www.joshualawrencechamberlain.com/bloodandfire.php.

Clausewitz, Carl von. *On War*. Boston, MA: Princeton University Press, 1976.

Connolly, Owen. *On War and Leadership*. Princeton, NJ: Princeton University Press, 2002.

Daddis, Gregory A. *No Sure Victory: Measuring U.S. Army Effectiveness and Progress in the Vietnam War*. New York, NY: Oxford University Press, 2011.

Dundas, Steve. Blog. "A Council of War: Meade and His Generals Decide to Stay and Fight at Gettysburg July 2nd 1863." Padre Steve's World. https://padresteve.com/2014/04/25/a-council-of-war-meade-and-his-generals-decide-to-stay-and-fight-at-gettysburg-july-2nd-1863/.

Eisenhower, Dwight D. "Remarks at the National Defense Executive Reserve Conference." November 14, 1957. Online by Gerhard Peters and John T. Woolley, *The American Presidency Project*. https://www.presidency.ucsb.edu/node/233951.

Fontenot, Gregory. *On Point: United States Army in Operation Iraqi Freedom*. Fort Leavenworth, KS: Combat Studies Institute Press, 2004.

Global Force Management Implementation Guidance. 2016. This document is classified and not releasable to the public.

Guidance for Employment of the Force. 2015. This document is classified and not releasable to the public.

Harrison, Gordon A. *Cross-Channel Attack*. Washington, DC: Center of Military History United States Army, 2002.

Hart, B. H. Liddell. *Strategy*. New York, NY: Signet Printing, 1974.

Heinl, Robert Debs Jr., ed. *Dictionary of Military and Naval Quotations*. Annapolis, MD: United States Naval Institute, 1967.

Historical Division. *Utah Beach to Cherbourg*. Washington, DC: History Division, Department of the Army, 1949.

Infantry in Battle. Washington, DC: The Infantry Journal Incorporated, 1939.

John E. Lange "Civilian-Military Cooperation and Humanitarian Assistance: Lessons from Rwanda." *Parameters*. Summer 1998, 106–122.

Joint Strategic Campaign Plan. This document is classified and not releasable to the public.

Jomini, Antoine Henri de. *Art of War*. Translated by G. H. Mendell and W. P. Craighill. Philadelphia, PA: J. B. Lippincott & Co., 1862. 17 November 2006. https://books.google.com/books?id=nZ4fAAAAMAAJ&printsec=frontcover&source=gbs_at b#v=onepage&q&f=false.

Lawrence, T. E. "The Evolution of a Revolt." *The Army Quarterly* Volume 1. London: William Clowes & Sons, Ltd. October 1920 and January 1921, 60.

McDonald, JoAnna M. *The Liberation of Pointe du Hoc: the 2nd U.S. Rangers at Normandy*. Redondo Beach, CA: Rank and File Publications, 2000.

Melcher, H. S. "The 20th Maine at Little Round Top." *Battles and Leaders of the Civil War* Vol. 3. Condensed from the "Lincoln County News." Waldoboro, Maine, March 13th, 1885. Available at http://www.joshualawrencechamberlain.com/20me7.php.

Moltke, Helmuth von. *Moltke's Military Works* Vol. 4, War Lessons, Part I. "Operative Preparations for Battle." Translated by Harry Bell. Fort Leavenworth, KS: Army Service Schools, 1916.

Montgomery, Bernard L. *The Memoirs of Field-Marshal Montgomery*. Cleveland, OH: The World Publishing Company, 1958.

National Defense Strategy of the United States. 2018. https://jdeis.js.mil/jdeis/jel/jel/other_pubs/nds2018.pdf.

National Military Strategy of the United States. 2015. https://jdeis.js.mil/jdeis/jel/jel/other_pubs/nms_2015.pdf.

National Security Strategy of the United States. 2017. https://jdeis.js.mil/jdeis/jel/jel/other_pubs/nss2017.pdf.

North Atlantic Treaty Organization Standardization Agreement 2199. *Command and Control of Allied Land Forces*.

Patton, George S. *War as I Knew It*. Boston, MA: Houghton Mifflin Company, 1947.

Porter, Horace. *Campaigning with Grant*. New York, NY: The Century Co., 1907.

Rommel, Erwin. *The Rommel Papers*. Edited by B. H. Liddell-Hart. New York, NY: Harcourt, Brace, and Company, 1953.

Sears, Stephen W. *Gettysburg*. New York, NY: Houghton Mifflin Company, 2003.

Sherman, William Tecumseh. *Memoirs of General W. T. Sherman*. New York, NY: Penguin Books, 2000.

Slim, Viscount William. *Defeat into Victory: Battling Japan in Burma and India, 1942–1945*. New York, NY: Copper Square Press, 2000.

Sun Tzu. *Sun Tzu on the Art of War*. Translated by Lionel Giles. London: Luzac and Co. 1910.

Unified Command Plan. 2011.

Washington, George. "From George Washington to Major General Stirling, 5 March 1780." *Founders Online*, National Archives. https://founders.archives.gov/documents/Washington/03-24-02-0525.

References

Widder, Werner. "Auftragstaktik and Innere Fuhrung: Trademarks of German Leadership." *Military Review*. September–October 2002 (English Edition), 3–9.

PRESCRIBED FORMS

This section contains no entries.

REFERENCED FORMS

Unless otherwise indicated, DA forms are available on the Army Publishing Directorate website: https://armypubs.army.mil/.

DA Form 2028. *Recommended Changes to Publications and Blank Forms*.

Index

Entries are by paragraph number.

A

actions, direct, coordinate, and synchronize, 2-29–2-33
   directing, 4-34–4-35
   sequencing, 2-75
   taking, 4-8
activities, assessment, 5-7–5-23
   battle rhythm, 1-82–1-83
   execution, 4-24–4-35
   operations process, 1-20–1-24
   planning, 2-85
   preparation, 3-12–3-33
adjustment decisions, 4-29
analysis, goals, 5-29
   span of control and, 2-21
Army command relationships, 2-22–2-24
Army design methodology, 2-89–2-91
   defined, 2-89
Army Ethic, operations process, 1-19
Army support relationships, 2-25–2-26
art of command, risk, 4-12
art of planning, 2-11–2-13
assess, commander, 1-49
   effectiveness, 1-49
assessment, activities, 5-7–5-23
   approach, 5-24–5-25
   characteristics, 5-7
   defined, 5-1
   directing action, 5-20–5-22
   during execution, 4-16–4-17
   effective, 5-32–5-39
   execution, 4-25–4-26
   formal, 5-5
   fundamentals, 5-1–5-6
   guides to, 5-32–5-39
   informal, 5-6
   measures, 5-15
   operations process and, 1-24
   resources, 5-5
   results from, 5-22
   working group, 5-35–5-36
assessment approach, 5-24–5-25
assessment plan, 5-33

developing, 5-26, 5-37
assessment process, 5-23–5-31
assumptions, develop, 2-135–2-138

B–C

basing, 2-81
battle rhythm, activities, 1-82
   characteristics, 1-82
   operations process and, 1-¬81–1-83
branch, defined, 2-39
brief, confirmation, 3-27
campaign plan, defined, 2-46
cause and effect, 5-39
CCIR, defined, 1-40
   influences on, 1-41
   visualization and, 1-40–1-43
center of gravity, 2-62–2-63
   decisive point and, 2-65
   defined, 2-62
   use of, 2-63
challenges, analysis, 5-18
   assessment and, 5-23, 5-39
   battle rhythm, 1-83
   developing solutions, 1-34, 2-15–2-17
   multinational operations, 1-29
   operations, 1-5–1-6
   planning, 2-132, 2-139–2-143
   plans, 2-33
   situations, 1-25
   thinking and, 1-69
changes, plans, 2-38
checks, conduct, 3-22
circumstances, adapt to, 2-34–2-40
collaboration, operations process, 1-61–1-62
   shared understanding, 1-33
collaborative planning, defined, 2-129
command, risk and, 1-78
command and control, defined, 1-11
   implementing, 1-15

commander's critical information requirement. See CCIR.
commander's estimate, situational understanding and, 1-57
commander's intent, 2-103–2-105
   defined, 1-38
commander's planning guidance, visualization and, 1-39
commander's visualization, 2-18
   characteristics of, 1-38–1-44
   defined, 1-34
   describe, 1-37
   execution, 4-25
commanders, battle rhythm and, 1-83
   center of gravity, 2-63
   considerations, 2-18, 4-36
   critical thinking, 1-66
   decisions, 4-11
   drive the operations process, 1-11–1-46
   focus, 4-18
   guidance, 2-60, 2-128
   initiative, 4-5
   involvement, 5-33
   operational art, 2-52
   planning, 2-87
   planning focus, 2-118
   preparation guides, 3-8
   responsibilities, 2-94, 2-116, 3-18, 4-34
   role, 1-31
   support to, 4-31
   support to, 2-64
   tools of, 2-30
communication, situational understanding and, 1-63
communications, liaison, 3-14
concept of operations, 2-106–2-109
   defined, 2-106
   operational art, 2-53
conceptual planning, 2-85
   Army design methodology, 2-91
conditions, creating, 4-8
   end state, 2-59–2-61
   evaluating, 5-19

Index

Entries are by paragraph number.

confirmation brief, defined, 3-27
control measure, 2-30–2-31
    defined, 2-30
coordination, liaison, 3-13–3-15
    plans, 2-120
    risk, 1-80
    targeting and, 1-77
creative thinking, 2-11
    apply, 1-65–1-70
    approach for, 1-68
    results from, 1-67
critical thinking, apply, 1-65–1-70
    results from, 1-66
culminating point, defined, 2-80

D

decision making, 4-27
decision point, defined, 2-37
decision support matrix, defined, 4-32
decision support template, 4-32
    defined, 2-38
decisions, types of execution, 4-27–4-30
decisive point, 2-64–2-67
    defined, 2-64
defeat mechanisms, applying, 2-55
    types, 2-54
depth, defined, 2-114
describe, commanders visualization, 1-37–1-44
detailed planning, 2-86
direct, commanders, 1-45
domains, multiple, 2-113
drive the operations process, principle, 1-31–1-49

E

efforts, prioritize, 2-18–2-28
end state, 2-105
    conditions and, 2-59–2-61, 2-68
    defined, 2-59
environment, effects of, 1-25
    mission command, 1-13
essential element of friendly information, defined, 1-44
ethical reasoning, thinking and, 1-69
evaluating, 5-13–5-19
    defined, 5-13
    results of, 5-21
events, forecast, 2-140

plan for, 2-34–2-40, 5-2
execution, activities, 4-24–4-35
    defined, 4-1
    fundamentals, 4-1–4-3
    guides, 4-4–4-17
    operations process, 1-23
    responsibilities, 4-18–4-23
execution decisions, 4-28
execution matrix, defined, 4-33

F

facts, determine, 2-135–2-138
feedback, 5-30
flexibility, defined, 2-117
    execution and, 4-3
    lack of, 2-142, 2-143
forces, integration, 3-21
    positioning, 3-7, 3-18
    protect, 3-10
    task organize, 2-18–2-28
framework, operation, 2-100
friendly force information requirement, defined, 1-43
functions, planning, 2-14–2-40
fundamentals, the operations process, 1-1–1-72

G–H–I

gambling, risk, 2-83
goal, operations process, 1-17
graphic control measure, defined, 2-109
guidance, assessment, 5-3
    planning, 1-39
guides, execution, 4-4–4-17
human endeavor, 1-4
indicator, defined, 5-14
    types, 5-15–5-17
influences, CCIR and, 1-41
information, analysis, 2-136, 5-28
    assessment, 5-9
    gather, 2-135
    mission variables, 1-52–1-53
    operational variables, 1-52–1-53information collection, 1-59, 5-11, 5-27
    defined, 1-74
    integrating process, 1-74–1-75
    preparation, 3-16
    results from, 1-75
    steps of, 1-74
information requirements, 5-10
    decisions on, 1-40
initiative, execution and, 4-4
    seize and retain, 4-5–4-12

inspections, conduct, 3-22
integrate, task-organize and, 3-6
    warfighting functions, 1-71
integrating processes, types of, 1-71–1-81
integration, 5-34–5-37
    forces, 4-23
intelligence, situational understanding and, 1-59–160
intelligence preparation of the battlefield. *See* IPB.
intelligence process, 1-60
interrelated options, planning and, 2-12–2-13
IPB, defined, 1-72
    described, 1-49
    integrating process, 1-72–1-73
    steps of, 1-73

J–K–L

joint, plan, 2-49
joint forces, support to, 1-10
key tasks, defined, 2-104
knowledge management, defined, 1-81
large-scale ground combat, characteristics, 1-6
lead, commanders, 1-46–1-48
leaders, considerations, 3-11
    qualities, 1-14
    responsibilities, 3-33, 5-4
    support to, 2-14
    tasks, 4-21
leadership, defined, 1-46
levels of warfare, defined, 2-41
    planning and, 2-41–2-51
    roles, 2-42
liaison, coordinate, 3-13–3-15
    situational understanding and, 1-63–1-64
line of effort, 2-68–2-72
    defined, 2-72
line of operations, 2-68–2-72
    categories of, 2-70
    defined, 2-69
    exterior, 2-71
    interior, 2-70
    use of, 2-68
location, commander, 1-48
logic, plan building and, 5-38

M

managing, time, 2-127
measure, assessment, 5-15–5-17
    graphic control, 2-109

Index

Entries are by paragraph number.

measure of effectiveness, defined, 5-17

measure of performance, defined, 5-15

mechanism, defeat, 2-54–2-55
    stability, 2-56

military action, rate, 2-73

military decision-making process, 2-92–2-94
    defined, 2-92

mission, defined, 2-101

mission command, 1-11–1-14
    approach, 1-1
    defined, 1-12
    principles of, 1-14
    risk and, 2-83

mission order, defined, 2-124
    details in, 2-126
    plans, 2-119–2-126

mission statement, 2-101–2-102
    elements, 2-101

mission variables, situational understanding and, 1-52–1-53

momentum, build and maintain, 4-13–4-15

monitoring, action and, 5-20
    assessment, 5-9–5-12
    defined, 5-9

movements, troop, 3-18

multinational operations, defined, 1-28

N–O

nature of operations, 1-1–1-8
    dynamic and uncertain, 1-5–1-6
    human endeavor, 1-4

nested concepts, defined, 2-108

network, preparation, 3-24

objective, decisive point and, 2-67
    principle of war, 1-7

operational approach, defined, 1-35
    operational art and, 2-53–2-56
    visualize and, 1-36

operational art, 2-52–2-84
    applying, 2-57–2-58
    defined, 2-52
    elements of, 2-57–2-84

operational concept, Army's, 1-9

operational framework, initial, 1-36

operational level, planning, 2-46–2-49

operational level of warfare, defined, 2-46

operational reach, defined, 2-79

operational variables, situational understanding and, 1-52–1-53

operations, adapting, 5-31
    command and control, 1-11
    context, 1-55
    human endeavors in, 1-4
    joint, 2-46
    nature of, 1-1–1-8
    political purpose of, 1-7–1-8
    purpose, 2-103
    security, 3-17

operations process, activities, 1-20–1-24
    assessment, 1-24
    by echelon, 1-26–1-27
    changing character, 1-25–1-27
    commander's role, 1-31–1-49
    defined, 1-15
    employment, 1-16
    framework, 1-15–1-29
    fundamentals of, 1-1–1-72
    goal, 1-17
    multinational operations, 1-28–1-29
    principles of, 1-30–1-70

opportunities, exploiting, 4-9
    momentum, 4-13
    risk and, 4-10

order, 2-109
    details, 4-38
    plan and, 2-5–2-7
    rapid decision-making and synchronization process, 4-37
    task organization and, 3-20

organization, communications, 3-15

P–Q

parallel planning, defined, 2-130

phase, change, 2-77
    defined, 2-76
    transitions and, 2-75–2-78

planners, considerations, 2-112

planning, activities, 2-85
    aids, 2-110
    assessment, 5-24
    challenges, 2-84, 2-132
    commander input, 2-87
    defined, 2-1
    detail, 2-141
    effective, 2-110–2-138
    functions of, 2-14–2-40
    fundamentals, 2-1–2-7
    goals, 2-51
    guidance, 1-20–1-21
    guides for, 2-110–2-138
    integrated, 2-85–2-99
    levels of warfare, 2-41–2-51
    liaison, 3-13
    methodologies, 2-88
    nesting, 2-50
    operations process, 1-21
    pitfalls, 2-139–2-143
    plans and, 2-1–2-5
    range, 2-134
    science and art of, 2-8–2-13
    techniques, 2-4
    time, 2-35
    value of, 2-6–2-7

planning horizon, defined, 2-133
    focus, 2-132–2-134

plans, adapting, 5-31
    assessment, 5-33, 5-37
    challenges to, 2-33
    combatant command campaign, 2-47
    complex, 2-120
    components of, 2-100–2-109
    developing, 2-119–2-126
    effects on, 4-24
    execution and, 4-2
    flexibility, 2-123, 4-35
    joint, 2-49
    long-range, 2-90
    modifying, 4-35
    order and, 2-5–2-7
    refine, 2-36, 3-32
    simple, 2-121–2-122
    support, 2-48
    task organization in, 2-20
    understanding, 3-4

policy, security, 2-44

positioning, commanders, 4-19
    forces and resources, 3-7

pre-operation, checks, 3-22

preparation, activities, 3-12–3-33
    defined, 3-1
    functions, 3-2–3-7
    fundamentals, 3-1–3-7
    guides to, 3-8–3-11
    liaison, 3-13
    operations process, 1-22
    prioritize, 3-9

presence, lead, 1-47

principle of war, simplicity, 2-119

priorities, determining, 2-18–2-28
    preparation efforts, 3-9

priority intelligence requirement, 1-25
    defined, 1-42

priority of support, defined, 2-28

problem, identifying, 2-90
    solving, 2-97

Index

Entries are by paragraph number.

R

rapid decision-making and synchronization process, 2-98, 4-36–4-38
readiness, assessment, 5-3
reasoning, ethical, 1-69
recommendations, creating, 1-56
reframing, Army design methodology, 2-91
rehearsal, defined, 3-28
relationships, command and support, 2-22–2-27
resources, 3-23
   adjusting, 4-30
   assessment, 5-5
   positioning, 3-7
risk, accept, 2-82, 4-10–4-12
   decision making, 4-28
   operational art, 2-82–2-84
risk management, defined, 1-79
   integrating processes, 1-78–1-80
role, commander, 1-18, 1-31
   leaders, 1-18
   operations process, 1-31–1-49
   staff, 1-18
   strategic, 1-9
running estimate, defined, 1-54
   situational understanding and, 1-54–1-58

S

science of planning, 2-8–2-10
security, force, 3-10
   policy, 2-44
security operations, preparation, 3-17
sequel, defined, 2-40
shared understanding, collaboration, 1-33
simplicity, 2-119
simultaneity, 2-112–2-113
   defined, 2-112
situational understanding, build and maintain, 1-50–1-64
   defined, 1-50
   preparation, 3-3
   tools to share, 1-51–1-64
situations, understanding, 2-15–2-17
solutions, developing, 2-15–2-17
span of control, 2-21
speed, 4-14

stability mechanisms, efforts from, 2-56
staff, considerations, 3-30, 5-24–5-25, 5-29
   coordination, 5-34
   decisive points and, 2-66
   execution, 4-22–4-23
   responsibilities, 2-104, 5-12, 5-21
   support to, 4-31
strategic level, planning, 2-43–2-45
strategic level of warfare, defined, 2-43
strategy, update, 2-45
structure, planning, 2-3–2-4
subordinates, support to, 2-131
   tasks, 4-20
success, exploit, 4-16–4-17
supervise, details, 3-33
supervise, preparation and, 3-11
support plan, 2-48
sustainment, preparation, 3-23
synchronization, 2-32, 2-115–2-116
   defined, 2-115
   initiative and, 4-6
   momentum and, 4-15
   plan, 2-47
synchronize, force, 1-71

T

tactical level, planning, 2-50–2-51
tactical level of warfare, defined, 2-50
targeting, defined, 1-76
   integrating processes, 1-76–1-77
   steps of, 1-76
task organization, defined, 2-20
   preparation, 3-19
task-organizing, defined, 2-19
   integrating and, 3-6
tasks, critical, 2-104, 3-5
tempo, 2-73–2-74
   defined, 2-73
   influenced by, 2-74
tenets of operations, defined, 2-111
tenets of unified land operations, planning aid, 2-111–2-117
terrain, management, 1-62, 3-25
   preparation, 3-26

thinking, detailed and conceptual, 2-88
   influences of, 1-65
   operations process and, 1-65–1-70
time, optimize for planning, 2-127–2-131
   planning, 2-35
   prioritize, 3-9
   transition, 3-30
tools, decision making, 4-31–4-33
   intellectual, 2-57
training, critical tasks, 3-5
   preparation, 3-21
transition, phasing and, 2-75–2-78
   plans to operations, 3-29–3-31
   priorities and, 2-78
   understanding, 3-4
troop leading procedures, 2-95–2-98
   defined, 2-95
   steps, 2-96
trust, liaison and, 1-64

U

uncertainty, planning, 2-14
   plans, 2-34
understand, 1-12–1-13
   commanders, 1-32–1-33
   preparation, 3-12
understanding, 2-85
   improve situational, 3-3
   opportunities, 4-9
   sharing, 1-51
   situations, 2-15–2-17
   subordinates, 3-27
   transition, 3-4
unified land operations, defined, 1-9
   operational concept, 1-9–1-10

V

variables, situational understanding and, 1-52–1-53
variance, 4-26
   decision making, 4-28
visualize, activities to, 1-35
   commanders, 1-34–1-36

W–X–Y–Z

warfare, levels of, 2-41–2-51
warfighting functions, integrating, 1-71
working group, assessment, 5-35–5-36

Made in the USA
Columbia, SC
29 July 2024